WHERE TO WATCH BIRDS IN MOROCCO

PATRICK AND
FÉDORA BERGIER

Illustrations by Mike Langman

Pelagic Publishing

Published by Pelagic Publishing
www.pelagicpublishing.com
PO Box 874, Exeter EX3 9BR, UK

Where to Watch Birds in Morocco
ISBN 978-1-78427-144-2

This book is a reprint of *The Prion Birdwatchers' Guide to Morocco* (2nd Edition, 2003, ISBN: 1871104092) published by Prion Ltd, Cley, UK.

Reprinted 2019.

British Library Cataloguing in Publication Data
A catalogue record for this book is available from the British Library.

Original design by Mike Hodgson.
Maps prepared by Steve O'Connor, Vellum Software.

Cover photographs by Patrick Bergier
Front cover: Cricket Warbler (*Spiloptila clamans*), an iconic species of South Morocco, taken at Oued Jenna. Back cover: the Cream-coloured Courser (*Cursorius cursor*) is widespread in desert Morocco, taken near Dakhla.

Printed and bound in India by Replika Press Pvt. Ltd.

FSC
www.fsc.org
MIX
Paper from responsible sources
FSC® C016779

The natural regions of Morocco

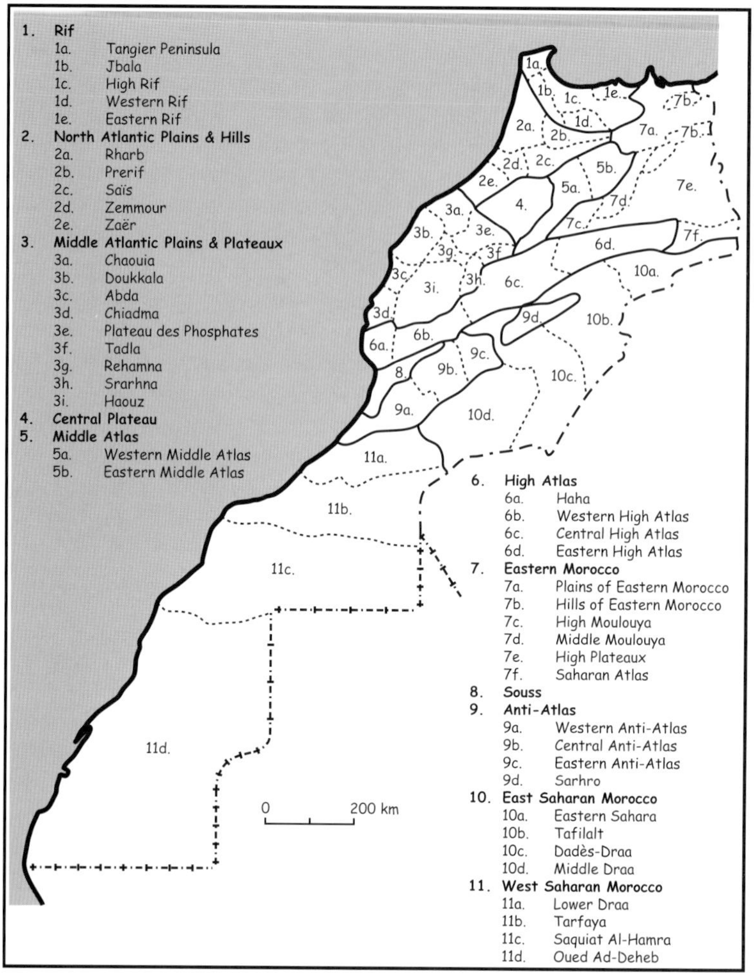

CONTENTS

The Sites

Appendices

ABOUT THE AUTHORS

Patrick Bergier was born in Provence, France, and graduated from a French Engineering School. He has a lifelong interest in the birds of the Mediterranean. He spent three years in Morocco working as an engineer in Rabat when he became involved with the study of Moroccan birds, and has subsequently visited the country a number of times. He has published many papers, his main interest being the distribution of birds and their breeding biology. Patrick is a founder member of Centrale Ornithologique Marocaine and a member of the Moroccan Rare Birds Committee (MRBC). He has co-authored *The Birds of Morocco* in the BOU checklist series.

His wife Fédora was born in Côte d'Azur, France, and graduated from a French Tourism School. She has travelled worldwide as a holiday tour leader and spent four years in Morocco where she gained an extensive knowledge of the customs, history and tourism resources of this country.

FOREWORD TO THE SECOND EDITION (2017 REPRINT)

Morocco has become an increasingly popular destination among birdwatchers. From a tourist standpoint, various regions have opened up widely; a lot of tracks have been asphalted, hotels have been built, and accommodation is no longer a problem in many areas.

The Moroccan people have learned to accept the presence of foreign visitors more readily and their welcome is more pleasant than ever. On the political side, the country is an oasis of quiet in a world of profound upheaval; the conflict with Front Polisario has calmed down and the Government is encouraging development – and tourism – in the Atlantic Sahara, which opens up great new sites for birdwatchers.

But these recent developments have also brought major disturbances to Morocco's bird life; in particular raptors have suffered dramatically from poisoning and habitat destruction, and for example Tawny Eagle and Dark Chanting Goshawk have disappeared in Morocco while Griffon Vulture and Lammergeier are on the verge of extinction.

We hope this edition will help the increasing numbers of visiting birdwatchers in their search for the Moroccan specialities and will help them appreciate this wonderful country.

Patrick and Fédora Bergier
e-mail : pbergier@yahoo.fr
website : www.go-south.org

ACKNOWLEDGEMENTS

This second edition wouldn't have been possible without the invaluable help of all those who provided us with their information, remarks, and reports, and the patience and good humour of those who came birding Morocco with us. Very many thanks to L. Albert, S. Albouy, G. Armstrong, S. Aulagnier, J.C. Barbraud, D. Barreau, P. Beaubrun, M. Bernade, Y. Bertault, L. Biron, N. Blake, C. Bowden, M. Carabella, P. Carter, S. Carter, B. Caula, S.P. Coyle, R. Cruse, F. Cuzin, H. Darmandieu, J.M. Daulne, B. Dawson, M. Deutsch, H. Dufourny, S. Durand, G. Elias, C. Engelhardt, E. Ferry, U. Fischer, F. Fornairon, A. Forsten, E. Franc, J.Y. Frémont, P. Geniez, F. Gérardin, M. Grundsten, M. Grussu, B. Guévorts, M. Hall, S. Hamidi, P. Huguenin, D. Jerez Abad, R. Jordan, A. Joris, A. Jouffray, E. Julliard, J.P. Julliard, H. Karhu, U. Karlsson, D. Lewis, A. Lezla, S. Lister, B. Mc Carthy, J.C. Paucod, F. Pianezza, J. Pineau, M. Pomarol I Clotet, T. Pons, C. Pouteau, R. Ramirez Espinar, N. Redman, F. Reeb, J.P. & J. Reitz, D.J. Robertson, P. Robin, H.P. Roche, A. Rocher, D. Rouable, K. Rousselon, P. Roux, V. Schollaert, B. Segerer, J. Ställberg, B. Théveny, F.L.L. Tombeur, J. Top, A.B. van den Berg, G. van Duin, J.D.R. Vernon, W.E. Waters, G. Willem, K. Wilson and J.P. and M.N. Zuanon, and apologies to anyone we have missed.

A very special thanks to Jacques Franchimont and Ahmed El Ghazi for their invaluable work at the GOMAC, our many days birding together, and our endless conversations on Moroccan birds, and to Michel Thévenot who has always supported us.

INTRODUCTION

Geography

Morocco is situated in the northwest corner of Africa, between c.21° and 36° North, and is basically an African country with a large Mediterranean region. Except for the extreme south, which is part of the Sahara, rains fall mainly during the cool season (October to April), and the summers are hot and dry; drought prevails in the Saharan region throughout the year.

The Country is divided into a number of geo-physical regions, and this influences the birds that are found there.

Mountains: five ranges are aligned along a northeast-southwest axis, and these include the Rif along the Mediterranean coast (2,456m, Jbel Tidighine), the Central Plateau (1,627m, Jbel Mtouzgane), the Middle Atlas (3,340m, Jbel Bou-Naceur), the High Atlas (4,167m, Jbel Toubkal) and the Anti-Atlas (3,304m, Jbel Siroua). Despite high human and animal pressure, forests of broad-leaved and coniferous trees are still widespread.

Atlantic Morocco includes rich agricultural plains and more barren and dry plateaux. Most of the original forest has been turned into matorral – or shrub-type vegetation – after intensive cutting and grazing, but some has been well preserved (such as the 'Forêt de la Mamora' with Cork Oak in the Rharb) however; large tracts of introduced eucalyptus have been planted. The Souss valley, with its unique Argan woodland, lies between the High and Anti-Atlas.

Eastern Morocco, except for the area adjacent to the Mediterranean coast, is dry with especially hot summers and cold winters; the High Plateaux rise to over 1,000m and are covered with a steppe type vegetation dominated by Wormwood *Artemisia herba-alba* and Halfa grass *Stipa tenacissima*.

Saharan Morocco lies to the south of the High and Anti-Atlas. In the eastern part, palm oases stretch along the rivers (Oueds) but the region is mainly large stony 'regs' (pebble desert) covered with *Hammada scoparia*; acacias thickets grow along the wadi bottoms. The drought is attenuated in a 15-200km wide strip along the Atlantic coast, which receives moisture from the ocean; the coast is mainly rocky.

The Birds

Morocco is a paradise for birdwatchers with its coasts, islands, wadis, plains, forests, mountains and deserts creating conditions that are often very different from those found in Europe. The richness of these habitats is clearly illustrated by the 452 species which have been recorded so far – see complete list in appendix – and 209 regularly breed there. Most (c.85%) breed north of the Atlas Mountains, because of the moister climate and more diversified habitats, and only about 35% breed in the Saharan region.

Every year, millions of migrants from western Europe pass through Morocco, mainly from late July to early November in the autumn and from March to May in the spring. Most, especially passerines, migrate by night or over-fly too high to be seen during the day; others, such as waders and gulls, often stop at wetlands to refuel, and provide unforgettable sights. The Straits of Gibraltar is famous for concentrating soaring birds, especially storks and raptors.

118 species are regular winter visitors, including 32 that are at the

southern limit of their wintering range. 117 other species have been recorded as accidental visitors, either from Asia (for example Great Knot), from Tropical Africa (Brown Booby and Lesser Flamingo), or from Northern America (Blue-winged Teal and Laughing Gull).

Birding hotspots

Several wetlands spread along the Atlantic coast are rightly famous for their migrant and wintering waders and gulls. These include the Lower Loukkos marshes, the merjas (lakes) of the Rharb, Merja Zerga, Lac de Sidi Bou-Rhaba, Sidi-Moussa-Oualidia lagoons, the Souss and Massa estuaries along the North coast, and Khnifiss lagoon and Dakhla Bay along the Saharan coast. The islets off Essaouira shelter a colony of Eleonora's Falcons.

The mountains shelter a rich avifauna; this can be seen best on the Plateau des Lacs in the Middle Atlas (Red-knobbed Coot, Levaillant's Woodpecker, etc.) and at Oukaïmeden in the High Atlas ('Atlas' Shore Lark, Alpine Accentor, Rock Sparrow, and Crimson-winged Finch).

Many desert-living species, including larks, wheatears, and sandgrouse, are widespread in the deserts of Morocco; others are more restricted in range, and birding hotspots include the temporary lake of Merzouga near Erfoud (waders and ducks in the desert!) bordered by the only large Moroccan sand dunes, the so-called Erg Chebbi (Desert Warbler, Brown-necked Raven, Desert Sparrow, etc.), Tagdilt, the Barrage Mansour-Eddahbi near Ouarzazate, the Draa valley, the oases of the Bani, and the West Sahara.

The Souss valley is famous for Dark Chanting Goshawk and Tawny Eagle, and the Straits of Gibraltar for the impressive raptor migration.

In this second edition, we attempt to describe a cross-section of sites covering most of the country's regions and main habitats. We include most of the better-known birding sites, and some less well known ones. The choice is to some extent a personal one but it is also dictated by factors such as access and travel restrictions, for example only the sites with current (2002) easy and regular access have been described in West Sahara.

There are, of course, hundreds of other sites worth exploring for Ruddy Shelduck, Marbled Duck, Black-shouldered Kite, Booted and Bonelli's Eagles, Lanner and Barbary Falcons, Double-spurred Francolin, Purple Swamphen, Houbara and Great Bustards, Cream-coloured Courser, 'Pharaoh' Eagle Owl (*ascalaphus*), Plain Swift, Blue-cheeked Bee-eater, Scrub and Tristram's Warblers, Fulvous Babbler, or Black-crowned Tchagra . . . so, when visiting Morocco, explore a few new sites yourself as well – you never know what you might discover.

The general naturalist

For the general naturalist there are many attractions other than birds. Minerals from the Middle and High Atlas will interest geologists; palaeontologists will find fossils (the belemnites and goniatites of the Tafilalt are famous); botanists will discover many unfamiliar species, among which are some marvellous orchids, and mammalogists, herpetologists and entomologists will also encounter numerous interesting species. At the end of this guide there are lists of all the birds, dragonflies, mammals, amphibians, reptiles, and orchids that have been found in Morocco to date.

A brief history

Man has inhabited Morocco since early times, as shown by pebble tools and rock carvings. After prehistoric times several races and cultures developed and mixed: Berbers, a white race whose origins are fairly poorly known but which still form the bulk of the Moroccan population today; Carthaginians; Romans (Volubilis near Meknès, was the first Roman city in Morocco and is well worth visiting). The Arabs invaded the country in the seventh century, bringing a new religion – Islam. Several dynasties followed one another throughout the centuries (Almohavides, Almohades, Merenides and Saadis) and the Alaouites reign at present. The French Protectorate began on 30 March 1912 and lasted until 1956. Today, the country is governed by H.M. Mohammed VI, who succeeded his father H.M. Hassan II in 1999.

The whole of Morocco, including West Sahara, covers 711,000km^2 (three times the size of Great Britain), but political problems with the Front Polisario over the West Sahara have not yet (2002) been solved completely, and regular access to the southern regions can be difficult.

The guide

The sites, which form the main part of this guide, have been chosen for their ornithological importance and also to allow the visiting birdwatcher to experience other aspects of Morocco and its culture. We have made every effort to make this second edition accurate and up-to-date (mid-2002), but we know that in a book of this nature some information will quickly become out-of-date. We would be grateful for any feedback, information, trip-reports, and corrections; please send them to the authors at: 4 Avenue Folco de Baroncelli, 13210 Saint Rémy de Provence, France (pbergier@yahoo.fr).

Red-rumped Wheatear

PRE-TOUR INFORMATION

Visas

A valid passport is needed to enter Morocco. No visa is required for European tourists staying less than three months. Other nationalities may be subject to different regulations and should contact their nearest Moroccan Embassy to ascertain the requirements for entry. No inoculations are required to gain entry to the country from Europe. Further information can be obtained from the Moroccan National Tourist Office, 205 Regent Street, London W1R 7DE (☎ 020 7437 0073) or the Moroccan Embassy, 49 Queen's Gate Gardens, London SW7 5NE (☎ 020 7581 5001 to 5004, e-mail: mail@sifamaldn.org)

Travel insurance

Comprehensive travel insurance is strongly recommended and should include full health cover and repatriation in case of a serious medical emergency.

Currency and exchange

The Moroccan currency is the Dirham (DH) which is divided into 100 centimes. In late 2002 the exchange rate was 16.65 DH to the pound (sterling), or 10.65 DH to the Euro approximately.

Customs

Passing through Customs is a relatively painless procedure. When entering and leaving the country, visitors have to fill in a form with information such as name, address, passport number, and profession.

As the customs regulations often change, it is advisable to check the current ones with the Consulates. In 2002, the import and export of Moroccan currency was prohibited but any amount of foreign currency could be brought in. All general goods could be imported duty-free (including camping equipment, a pair of binoculars, a telescope, and a camera), provided they correspond to a regular tourist activity. In addition, each adult could import one bottle of spirits and one of wine (or two bottles of wine) and 200 cigarettes. Professional cameras, video-recorders and other expensive electronic equipment may be entered on your passport by Customs officials and you will be charged a 100% duty if you leave Morocco without them.

Photography and birdwatching

Even though tourists and birdwatchers are now very well accepted throughout Morocco, do not use camera, binoculars or telescope in 'sensitive' areas, such as near a military zone. Bring plenty of film. If you run out more may be purchased in tourist areas or in the main cities, but prices are often high. Store your optical equipment in a special bag to keep it free from dust and sheltered from the sun.

Planning your trip

One of the best times for birdwatching, from March to June, also corresponds to the busiest tourist period and it is wise to book your flights, car hire and hotels (especially in the main tourist areas such as Marrakech or Agadir) well in advance.

Websites

Many websites will provide you with information when preparing for your trip, and their numbers increase every day. Here are just some of the general ones:
http://www.travelnotes.org/Africa/morocco.htm,
http://www.webguidemorocco.com, http://www.morocco.com and

http://goafrica.about.com/travel/goafrica/cs/morocco (in English), and http://www.tourisme-marocain.com/, http://www.bestofmaroc.com/ and http://www.maroculture.com (in French).

Also take a look at the authors' website (http://www.ifrance.com/go-south) mainly devoted to birds and nature in Morocco.

In case of Emergency

Useful addresses in case of emergency are: the British Embassy, 17 Boulevard de la Tour Hassan – Rabat (☎ 037.72.09.05 and 06) or one of the British General Consulates: 60 Boulevard d'Anfa – Casablanca (☎ 022.22.16.53), 9 rue d'Amérique du Sud – Tangier (☎ 039.93.58.95), and Beach Club Hotel, Chemin de l'Oued Souss – Agadir (☎ 048.84.43.43).

Other Embassies are also located at Rabat: France (3 Zankat Sakhnoun, Agdal. ☎ 037.68.97.00), Germany (n°7, rue Madnine. ☎ 037.70.96.62), The Netherlands (40 rue de Tunis. ☎ 037.72.67.80), United States (2 Avenue Marrakech. ☎ 037.76.22.65).

TRAVEL INFORMATION

Travelling to Morocco

By air

Morocco has 11 major airports, and five cities regularly receive international flights; Tangier, Casablanca, Marrakech, Ouarzazate, and Agadir. If visiting the north of the country it is better to land at Tangier or Casablanca, for visits to the south Agadir and Ouarzazate are best, while Marrakech occupies a central position for visiting the main birdwatching sites.

All the airports are fairly near the city centres. They have all the usual facilities including banks and car hire. Be sure to reconfirm return flights at least 72 hours in advance.

By land

Allow for a four-day journey if driving down to Morocco from the UK, through France, then Spain. The shortest and less crowded route is via Paris, the highway A10 down to Bayonne, then Madrid, Cordoba, Seville, and Cadiz. Take a car ferry at Algesiras (more ferries than from Tarifa or Gibraltar) to either Ceuta (a Spanish enclave on the Mediterranean coast of Africa) or Tangier, crossings that take only some 30 to 60 minutes.

There are regular Eurolines coach services between London and most other major European cities to Morocco, and these can be convenient for those with a lot of material to bring and who are intending to rent a car once in Morocco; but this is also a fairly long journey and it is likely to take three days or so (http://www.eurolines.com).

Finally, you can also get there by train from London (in some three days) or from most of the European capitals (for example, daily departures from Paris, 26 hours to Algesiras; http://www.sncf.fr/).

By sea

A good, quieter and safer alternative is to travel part of the journey by sea, taking a car ferry either in Spain at Almeria or Malaga (to Melilla, another Spanish town in Africa – 6h 30min to 7h 30min) or in France at Sète (to Tangier all year round and to Nador in summer only – 36h). Advance booking is essential, especially if visiting at Easter or in summer time; contact your travel agent or one of the specialists, e.g. Transmediterranea (http://www.trasmediterranea.es/) or Euro Mer (http://www.euromer.net/).

Travelling within Morocco

By car

The most convenient way of touring Morocco is by hiring a car. The driver must be over 21 and have been in possession of a driving licence for at least a year. Most of the large car-hire companies are found in the main cities, but there are also local ones which are sometimes less expensive. Small models such as the Fiat Uno or Peugeot 205 are readily available (€280 – 340 per week in 2002, including unlimited mileage and comprehensive insurance, but only if you stay on the asphalted roads). Larger models are also available. Most of the companies also offer four-wheel drives, minibuses and Land Rovers with eight seats, driven by a local driver who can act as a guide, interpreter, and cook. This last suggestion might be of interest to a group of birdwatchers visiting the country for the first time, or for those who intend to go birding in more remote areas.

Most of the Moroccan asphalted roads are of good quality. This network, which covers about 60,000km, is complemented by a long network of dirt roads. The quality of the small connecting roads and tracks varies greatly. Some of them, although shown on maps, can be impractical during certain months of the year, or completely abandoned. Take this into account if intending to do a long journey off the main roads. If exploring remote areas, take two cars as it is no fun to be alone in a remote area, with a broken down car!

When travelling in West Sahara, stay on the asphalted roads and avoid the tracks or driving or walking in the wild, as mines were spread over large areas during the conflict with Front Polisario, and still remain there.

When driving, always bear in mind that anything can happen on Moroccan roads! Watch out for children when driving through villages, and for motor cycles and cattle by the wayside. Try to avoid driving at night as some vehicles have no lights and pedestrians often walk on the roads.

Petrol stations can be found all over Morocco, even in the most remote areas. Petrol cost about 9 DH (€0.9) per litre in 2002, but in West Sahara south of Tantan is only half this as there is no tax.

By coach

A cheap way to travel in Morocco is by coach. Coach lines go everywhere in the whole country; their prices are fixed and attractive, and their comfort has recently been improved, with air-conditioning for some companies (such as Satas or Supratours, which also cover the deep south down to Dakhla). Improvements have also been made to timetables. Nevertheless, the oldest and cheapest company, the C.T.M. (Compagnie des Transports Marocains), is often crowded (especially during the mornings in the south, due to the heat), and its timetable is unreliable.

By train

The railway is fairly well developed in the north of the country, with two main lines: Tangier-Rabat-Casablanca-Marrakech and Casablanca-Rabat-Fès-Oujda. The lines have recently been improved, and the train is now a very pleasant and cheap way to travel between these cities. More information and timetables are available on the Office National des Chemins de Fer (ONCF) website (http://www.oncf.org.ma/).

Taxis

The taxi system comprises 'small taxis' (individual) and 'large taxis' (collective). Individual taxis only operate inside the towns; they are cheap – even if the price tends to be higher for foreign tourists than for local people. They are numerous, all painted in the same colour in one town, and are much more convenient than city buses. If there is no meter in the taxi, ask how much the fare will be before setting off. The 'large taxis', generally big old cars, take up to six people, and are used for trips from one town to another. They leave only when they are full and it is possible to bargain to get a good price.

Hitching Hitch-hiking is not a common practice and a share in expenses may be asked for, however it can be useful in remote areas.

By air Air travel within the country will save those travellers with limited time a few day's travel; the National company Royal Air Maroc (http://www.royalairmaroc.com/) operates a regular service from Casablanca airport to Agadir, Al Hoceima, Dakhla, Fès, Layoun, Marrakech, Ouarzazate, Oujda, Rabat, Tangier and Tétouan, and this can be of great help when visiting remote sites, especially those along the West Sahara Atlantic coast.

Crested Coot

STAYING IN MOROCCO

Accommodation

Many tourists visit Morocco and there is a wide variety of accommodation to suit all pockets, ranging from campsites to 5-star hotels. The higher quality hotels are concentrated mainly in the larger towns, such as Marrakech, Agadir, Ouarzazate, Tangier, Casablanca, Rabat, Meknès, and Fès. Elsewhere, they are few and far between. The availability of accommodation has increased greatly recently and the country is now very well equipped with modest or medium-range hotels (equivalent to 1- or 2-star) and inns. They are usually well kept and prices are attractive (100 to 200 DH – €10-20 – per room for two persons) and so they generally suit most birdwatchers as they offer adequate comfort. Most of them also have restaurants.

Morocco also has a lot of campsites in the tourist areas, and some of them are pleasant, well situated, cheap, and clean. In the non-tourist areas there are very few designated campsites. In these instances it is a good idea to get in touch with the local police, the chief of the village or, failing that, a local person to ask where to camp. Moroccan tea or food will sometimes be offered – don't refuse as you may offend (the custom is to offer three glasses of tea). It is also considered polite to give a present when leaving the site.

Food

Moroccan cooking is said to be one of the best in the world with some of the most delicious oriental dishes. The food is spicy but rarely very hot, and comprises such dishes as couscous, tajines, soups, kebabs, and pastries. Many different dishes can be found in the numerous restaurants in any town or village.

Couscous is the national dish, combining semolina with a large variety of vegetables cooked with mutton or chicken. Tajines are usually cooked and served in a special round baked-clay dish (called a tajine) and are the equivalent of stews. They can be made with mutton, chicken or fish, with vegetables or almonds, dried plums or olives. Harira is the most popular soup, made with a little meat and lentils – the best ones are often made and sold by women in small village markets. Pastilla is the masterpiece of Moroccan cooking. It is composed of many very thin and crusty pancakes covering a delicious pigeon and almond stuffing. Kebabs are one of the most common dishes. Small restaurants at the main coach stops offer kebabs and glasses of tea but take care to avoid any with chopped meat as these may have been reheated. Moroccan pastry is of oriental origin and is based on almond paste, sometimes highly coloured, and honey. It is also found in other Islamic Mediterranean countries. The most famous is called 'corne de gazelle'.

In 2002, a typical meal, including salad or soup, tajine and dessert cost 60-90 DH (€6-9).

While on the coast don't miss the opportunity to sample the different types of Moroccan seafood. Try for example the oysters from Oualidia (page 44) or the fish from the Essaouira harbour (page 49).

Tea-houses are quite common in the main towns and both oriental and European pastries are normally available.

In addition to typical Moroccan food, international cuisine, largely influenced by the French, can be found mainly in the restaurants of the 3- to 5-star hotels.

In any town or village, ample supplies can be found in the grocers, butchers, and bakers. Local people prefer to go shopping at the open-air markets, called 'souks', which have a unique atmosphere.

Drink

Cafés offering soft drinks, coffee, and tea are found everywhere throughout the country. Most alcoholic drinks can only be bought in hotels and in a few stores in main cities. Wine is only available in the tourist restaurants and in some stores in the biggest cities. Beer is more commonly available, but Moroccan people generally drink water or mineral water (Oulmes, Sidi Harazem, Sidi Ali), soft drinks, coffee, and of course, the famous tea with mint.

Banks

Banks are numerous in all the main towns and even in the smaller ones. They generally open from 08:30 to 11:30 and 15:00 to 17:00. During the Ramadan period (an Islamic month of fasting currently in the winter), the duration of opening is shortened, from 09:30 to 14:00. During weekends currency can be exchanged at the reception desks of most main hotels, and Dirhams can be obtained from cash machines at some banks (in the major tourist cities) using a credit card.

While traveller's cheques and the main international credit cards (American Express/MasterCard/EuroCard and above all Visa) offer a safe way to carry money, their use is less widespread than in Europe, but is gradually increasing. They will be useful for paying expenses in the main hotels and big souvenir stores, and to get Dirhams in the tourist areas and the business districts. Foreign currency is more easily changed in almost all the banks.

Holidays

Working life in Morocco is based on the Gregorian calendar. Weekends comprise Saturdays and Sundays; the opposite of what is found in some other Muslim countries, Friday is a work day and shops, administrative services and banks are open but tend to close earlier (or to extend the noon break) to allow people to go praying.

National holidays are 1 and 11 January, 1 May, 30 July, 14, 20 and 21 August, and 6 and 18 November.

Religious life follows the Muslim calendar which began on 16 July 622 (the day when Mahomet left Mecca for Medina); the hegiran year also has 12 months but is based on the moon rhythm and is shorter than the solar year: the dates of the month of Ramadan and the other religious holidays vary from year to year (they go backwards by some ten days each year).

Religious holidays are the Aïd el Adha (mid February in 2003), the First of the Hegiran Year (early March in 2003), Mouloud (mid May in 2003), first day of Ramadan (late October in 2003), and Aïd el Sghir (late November in 2003).

Staying connected to the rest of the world

The major Post Offices in the big cities generally open between 08:30 to 18:30, but most others close between 12:00 and 14:00; they are generally crowded, and it is wise to buy stamps when buying post cards.

The telephone network has been greatly improved, and 'Téléboutiques' are found in every town and village; these small, blue shops allow easy local or international phone calls for decent prices. Since autumn 2000, Moroccan phone numbers have nine digits (the first one is 0) and these nine digits have to be dialled for each local and national calls. To call Morocco from Europe or the rest of the World, just dial the international code for Morocco (212) then the last eight digits of the number (i.e. do not dial the first 0). Cell phone numbers begin by 06.

Cybercafés have also developed greatly and connecting to the Internet is now easy and cheap (12 to 25 DH per hour in mid-2002), even if connections tend to be slow sometimes, especially in the late afternoon.

Tips

Giving money and goods is a requirement of Islam and this is a natural function for Moroccans. Tourists are often asked to give something but we encourage visitors to give cautiously; in particular, giving to children encourages begging (see The Tourist Code below). On the other hand, it is customary to give some DH to the barman, the waiter ... in other words to anyone who provides a service. Consider giving one or two DH to the petrol pump assistant when refuelling and 4 or 5 DH to a 'car-keeper' when parking (10 DH for a night – this will reduce the chance of robbery, but still don't leave anything conspicuous inside your car).

Bargaining

Moroccans expect people to bargain for whatever they are buying. Bargaining is entirely natural in Morocco and visitors should be prepared to do so. There are of course no hard rules but there are a few general points to keep in mind. First, bargain only for something you really want to buy and only mention a price if you are willing to pay that amount. Second, bargaining takes time so be ready to spend a couple of hours talking and drinking tea in order to get a 'good price' for something important, such as a carpet. Third, don't pay attention to the initial price; the final price may be completely different, but there is no specific 'reduction ratio'. Finally, a 'good price' should be what you were expecting to pay and what the vendor was expecting to receive.

The Tourist Code *(from an original idea of Discover Ltd.)*

As a guest, respect local traditions, protect local cultures, maintain local pride.

When taking photographs, respect privacy – ask permission and use restraint.

Respect religious and cultural places – preserve what you have come to see, never touch or remove religious objects.

Giving to children encourages begging – a donation to a project, health centre, or school is a more constructive way to help.

You will be accepted and welcomed if you follow local customs – use only your right hand for eating and greeting. It is polite to use both hands when giving or receiving gifts.

Respect for local etiquette earns you respect – loose, light-weight clothes are preferable to revealing shorts, skimpy tops, and tight fitting action wear. Holding hands or kissing in public are disliked by local people.

Visitors who value local traditions encourage local pride and maintain local cultures – please help local people gain a realistic view of life in Western countries.

Be patient, friendly and sensitive!
Remember – you are a guest!

African Marsh Owl

CLIMATE AND CLOTHING

Climate

Morocco has been described as a cold country where the sunshine is hot. North of the Sahara, the Moroccan climate is characterised by mild winters, hot summers, and generally modest rainfall, concentrated from autumn to spring (especially in November and February-March). As a whole, this climate has been defined as 'Mediterranean' but in fact it varies widely – from the Atlantic ocean to the Algerian border, from the Mediterranean coast to south of the Atlas mountains, and from sea level to the top of the High Atlas.

The following map shows the distribution of the main bio-climatic areas of Northern Morocco (1: damp, sub-damp and high mountain; 2: semi-arid; 3: arid; 4: Saharan).

The bio-climatic areas of Northern Morocco

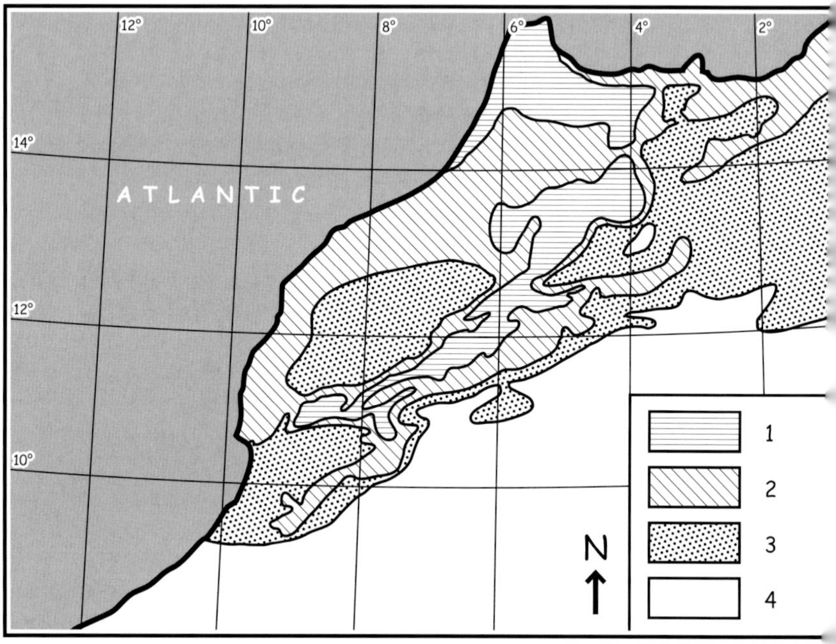

Temperature and rainfall vary widely across the country. The rainfall averages range from 25mm a year in the Draa valley near Zagora to 2,000mm in the Rif. Temperature differences between summer and winter in a single place can vary by 25-35°C. Moreover the Moroccan climate varies widely from year to year.

Average daily maximum temperatures (in degrees Celsius) for a selection of places and months are presented in the following table.

	January	April	July	October
Tangier (site '*The Straits of Gibraltar*')	15.4	19.2	26.4	22.1
Essaouira (site '*Essaouira*')	18.2	19.8	21.7	21.9
Ifrane (site '*The region of the Lakes*')	8.5	15.7	30.6	18.7
Marrakech (site '*N'Fiss reservoir*')	18.1	25.7	37.8	28.1
Agadir (site '*The National Park of Souss-Massa*')	20.3	23.3	26.4	25.9
Ouarzazate (site '*Ouarzazate*')	17.3	26.9	39.4	27.0
Zagora (site '*The Draa Valley*')	21.2	30.2	43.6	30.6

Clothing

The most suitable clothing will depend upon the time of year and location. Generally casual clothes and good shoes for walking will suffice. Neutral colours are more suitable than bright ones when birdwatching. In summer, avoid synthetic materials and choose linen or cotton. Don't forget a hat, a swimsuit and sunglasses. Even during the hot months, bring a pullover for the evenings and, if intending to take a trip in the mountains, a warm anorak. From late autumn to spring, take warm clothes and a waterproof jacket.

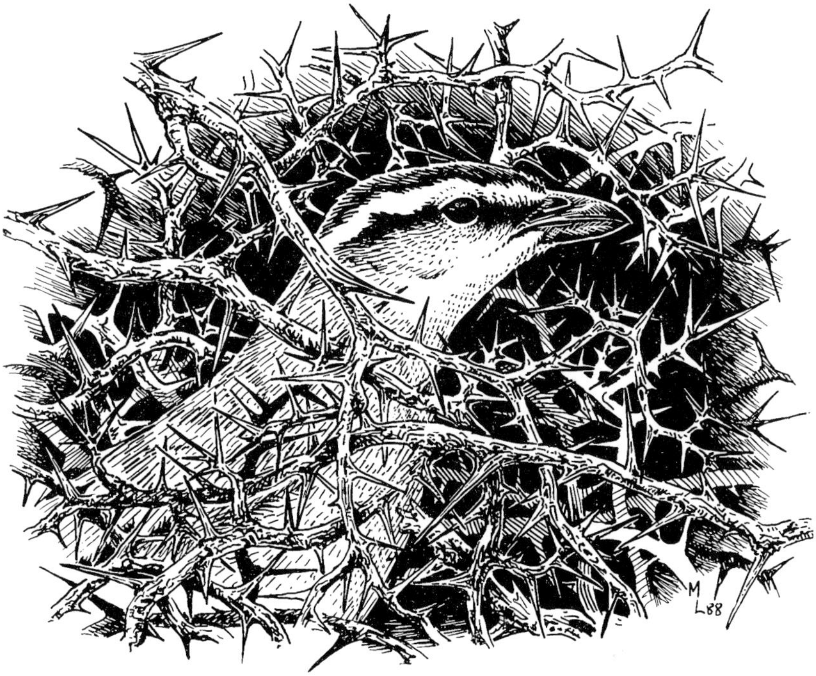

Black-headed Bush Shrike

HEALTH AND MEDICAL FACILITIES

There is very little chance of catching a serious disease in Morocco. However, some precautions should be taken to ensure a pleasant stay. Firstly, be sure to be up to date with all standard vaccinations. Although no specific inoculations are required for entry it is wise to be immunised against cholera, typhoid, polio, and tetanus.

The most frequent problem that European tourists encounter in Morocco is gastro-enteritis. Although often mild it can still be a great nuisance. It is difficult to avoid as it is often due to a change in diet. Following some simple rules will help to prevent a bad attack. Eat well-cooked meat; wash vegetables and fruits; don't drink too much fruit juice, and try to drink only bottled drinks and bottled or mineral water, especially outside the main towns. For those intending to spend a long period outside towns or tourist sites, it is a good idea to take tablets for water sterilisation. The symptoms of common gastro-enterisis can be controlled by tablets such as Imodium and Ercefuryl (available in some Moroccan pharmacies). With severe or persistent symptoms a local doctor should be consulted.

If bitten by a dog or a snake (fortunately extremely rare) seek treatment by a doctor as quickly as possible. Anti-snake bite serums are available, but they are difficult and inconvenient to keep.

There is no fatal scorpion in northern Morocco, but their bites can sometimes be painful and give some trouble; don't hesitate to take medical advice. In the south, avoid walking barefoot in the marshes and bathing in the wadis because of the risk of bilharziosis.

Lastly, be careful to avoid sunburn. Bring adequate clothes (especially long-sleeved shirts and a broad-rimmed hat), sunglasses, and protective cream. Mosquito repellent will also be useful.

Every main town has good hospitals and clinics and there are also numerous doctors (general practitioners and specialists) and pharmacies. In villages and small towns, medical facilities are not as comprehensive and often only a dispensary exists. Be sure to check that your travel insurance covers medical costs, including repatriation in case of severe illness or accident.

MAPS, BOOKS AND OTHER INFORMATION

Maps

Since the acquisition of the new Saharan Provinces, Morocco stretches as far south as 21°N and the new maps show the entire country.

The maps most commonly found are at a scale of 1:1,000,000 and have been made by the French Institut Géographique National (http://www.ign.fr/), Blay Foldex (http://www.blayfoldex.com/) or Hallwag (http://www.hallwag.ch/). The Hildebrand's (http://www.hildebrands.de/) is at a scale of 1:900,000, the Geo Center is 1:800,000. One of the best is certainly the Michelin (no. 959), scale 1:1,000,000, which is updated every two years.

Most of these maps are available from Stanfords International Map Centre, 12-14 Long Acre, Covent Garden, London WC2E 9LP (http://www.stanfords.co.uk/) and from every FNAC (http://www.fnac.com/) in France.

In Morocco only a few maps are normally available, so it is wise to bring one from home.

An excellent series of maps (scales 1:100,000 and 1:50,000) which are most useful for birdwatchers are produced by the 'Division de la cartographie' (Avenue Hassan II, km 4, Rabat. ☎ 037.29.50.34, or 51.17 and 55.48). Those planning a long visit, undertaking a detailed study in a special area, or taking a trip off the main roads, should stop at Rabat in order to buy the relevant maps. A few of these are available from Stanfords in London.

General Guides

A number of general travel guides are available for Morocco, for example the 'Lonely Planet' (http://www.lonelyplanet.com/) or 'Footprint Handbook' (http://www.footprintbooks.com/) and each has its own advantages, but the authors' prefer the 'The Rough Guide' (http://www.roughguides.com/) in English and the 'Guide du Routard' (http://www.routard.com/) in French.

Field Guides

Several field guides cover Morocco. One of the most comprehensive is the 'Collins Bird Guide' by Svensson, Mullarney, Zetterstom and Grant and this is the most useful in the field. Other good recent books include the 'Birds of Europe with North Africa and the Middle East' by Jonsson and the 'Handbook of Bird Identification' by Beaman and Madge, though the latter is too large and heavy for field use.

The recent editions of the 'Birds of Britain and Europe with the Middle East and North Africa' by Heinzel, Fitter and Parslow may also be useful. The 'Birds of the Middle East and North Africa' by Hollom, Porter, Christensen, and Willis covers Moroccan specialities but needs to be used in conjunction with 'A Field Guide to the Birds of Europe' by Peterson, Mountfort and Hollom. Other books and papers which may prove helpful are listed in the bibliography at the end of this guide.

Audio Guides

In 2000 the Société d'Etudes Ornithologiques de France (S.E.O.F., 55 rue de Buffon, 75005 Paris – France) published a series of four compact discs on the songs of the birds of the Sahara, North Africa and the Atlantic islands of the Western Palearctic. They cover 423 species including all those likely to be found in Morocco. They are very good value for money and are well worth taking on a trip.

A number of websites have collections of bird songs, including some North African species.

Trip reports

Trip reports are an invaluable source of information when preparing for a visit as they contain up-to date information. Some organisations have specialised in gathering such reports and can supply them at a small per page fee. Two of the best known are 'Birdtours' (http://www.birdtours.co.uk/) and the 'Foreign Bird Reports and Information Service' (FBRIS. 6 Skipton Crescent, Berkeley Pendesham, Worcester. WR4 0LG – UK. http://www.ukbishosting.co.uk/fbris/).

The authors' website (http://www.ifrance.com/go-south/web/tripreports.htm) also has a selection of new reports that can be freely downloaded.

WHEN TO GO

In the northern half of Morocco, between 28°N and 36°N and 1°W and 12°W, the biological seasons are the same as in Western Europe.

From an ornithological point of view, the period from **March** to **May** is probably the most interesting as a wide variety of species, migrants as well as residents, are present. This is the most pleasant time of the year as after the winter and early spring rains the weather is generally mild and the sun shines most of the time.

Among the most conspicuous migrants are raptors and waders. Most of the Lesser Kestrels pass through in March, Black Kites, Short-toed Eagles and Hen Harriers in March-April, and Honey-buzzards in April. Ospreys, Booted Eagles and Hobbies pass through anytime from March to May, and the migration of Montagu's Harriers begins at the end of March and goes on until the beginning of May.

Most waders such as Black-winged Stilts, Grey Plovers, Red Knots, Sanderlings, Little Stints, Curlew Sandpipers, Dunlins, Ruffs, Curlews, Greenshanks, and Wood and Common Sandpipers migrate from March to May. Green Sandpipers are mainly seen from mid-March to mid-April. The first Collared Pratincoles arrive in mid-March.

On the Atlantic coast, most species of gulls and terns migrate throughout the spring. Audouin's Gull moves from February to mid-April, Black Tern from late March to the beginning of May and Whiskered and White-winged Black Terns from the beginning of March to the beginning of May.

Of the migrants that breed in Morocco, Red-rumped Swallow, Tawny Pipit, Desert Wheatear, Subalpine and Western Bonelli's Warblers, and Woodchat Shrike arrive in early March. From mid-March European Bee-eater, Short-toed Lark, Nightingale, Northern Wheatear (Moroccan race), Black-eared Wheatear, and Spectacled and Orphean Warblers can be seen. Blue-cheeked Bee-eater, Roller, and Western Olivaceous Warbler begin to appear from late March. From the beginning of April Rufous Bush Robin, Melodious Warbler, Spotted Flycatcher, and Golden Oriole arrive.

In the plains north of the Atlas Mountains, breeding generally begins in March and goes on until June. In semi-desert and desert areas, the season is approximately the same. Early nests have been found in January (for example Hoopoe Lark) during favourable (wet) years, although even then laying can occur much later. Sandgrouse breed mainly in June, when the weather is becoming very hot.

In **June** the climate begins to become uncomfortably hot, except near the coast or in the mountains. At Oukaïmeden for example (page 67), breeding is in full swing (Horned Larks, Tawny Pipits, Black and Moussier's Redstarts, Northern Wheatears, and Rock Sparrows).

In **July** and **August** the heat may be unbearable away from the coast and the high mountains. Breeding is over, and some species have already begun their autumn migration by August (for example Honey-buzzard, Black Kite, Montagu's Harrier, and Booted Eagle).

In **September** many species begin to migrate and the first wintering birds are generally noted from the end of this month, or in **October**. During this period, the best sites to visit are located on the Atlantic coast but a visit to the Straits area (page 23) to observe the unforgettable raptor migration is well worthwhile.

In **November** most of the wintering birds have arrived in the country. This month is generally one of the wettest of the year and all the wetlands along the Atlantic coast are worth visiting. With a little luck, some rarities will be discovered among the more common birds.

Even though the temperatures are decreasing, **December** and **January** are generally more agreeable, with less humidity and more sunshine. Christmas and New Year are particularly pleasant along the southern Atlantic coast. If you intend to take a trip at this time, try to spend at least a day at Massa (page 84). Besides the thousands of coots and ducks, the first spring migrants of such species as Great Spotted Cuckoo, Hoopoe, Pallid Swift, Barn and Red-rumped Swallows should be seen.

In the semi-desert and desert areas, the nights are fairly cold (water can freeze in the palm-groves) but the days are generally pleasant. Take the opportunity to visit such sites as the Ouarzazate Barrage and the Draa valley (page 113-116) or the Tafilalt (page 106), with the famous daya of Merzouga – a temporary lake in the open desert which attracts hundreds, sometimes thousands, of birds during wet years.

During this period mountains are covered with snow, and several roads and passes are blocked. Birdwatching becomes difficult there, except in places such as the Ifrane region (page 54) or Oukaïmeden (page 67), two spots where ski resorts have been built and the roads are kept open.

In **January** and the beginning of **February** the largest concentrations of White Storks may be seen. Several hundred birds often gather (for example 700 in February 1981, on the refuse dump of Marrakech city). In February the rains begin again, and allow the vegetation to grow. The spring migration intensifies and by the end of the month the first breeding begins.

Throughout the year a birdwatcher will always find something interesting in Morocco, whether in the plains, wetlands, mountains or deserts. Always bear in mind that, depending on the time of year, some places will be more productive than others.

INTRODUCTION TO THE SITE INFORMATION

The places described include not only specific sites such as the mouth of the Souss and Massa rivers, the lake of Sidi Bou-Rhaba, and the biological reserve of Moulay Bousselham, but also wider areas. Generally, these areas contain smaller densities of birds and have been included because of their important ecological interest and/or the beauty of the landscape.

Overseas birdwatchers can become familiar with the typical Moroccan fauna and flora by visiting these habitats: desert and oases (Tafilalt, Draa valley, Jbel Bani); semi-desert steppes with halfa-grass (upper valley of the Oued Moulouya, Zeïda); the cork-oak and Sandarac woods (Zaër); the cedar-woods and lakes of the Middle Atlas (the region of the lakes); the valleys and mountains of the High Atlas (Tizi-n-Test and Tizi-n-Tichka roads, Imlil, Oukaïmeden, the valleys and gorges of Oueds Todra and Dadès); the Argana forest (Souss); the Juniper woods (Essaouira).

The name of the site is followed by the region where it is located, and thus can be easily cross-referenced with the following map.

The natural regions of Morocco

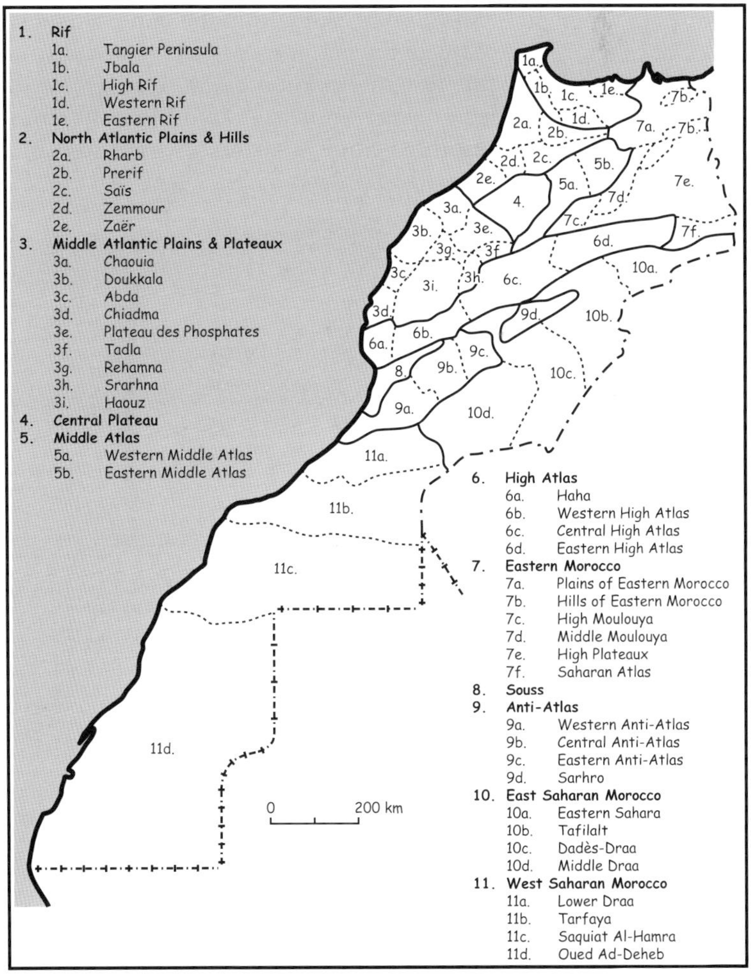

The sections on accommodation only list the hotels that the authors can recommend at the time of writing this 2nd edition (mid-2002); tourism is developing rapidly in Morocco, and accommodation availability, condition, and price constantly change. For regular updates, consult the site sheets (in French) available through the authors' website (http://www.ifrance.com/go-south/oursites/oursites.htm), or the latest edition of one of the several general tourist guides that may be found in most bookshops.

We have tried to give the approximate price of the hotels that we suggest in the site descriptions, according the following, for double rooms:

- ❶ upper category, more than 400 DH (€40)
- ❷ middle category, 200-400 DH (€20-40)
- ❸ budget, less than 200 DH (€20)

Remember that these tariffs were valid in 2002. Always ask to look at a room before committing to it, and remember that it is also possible to bargain in some places!

The sections on birds do not list all of the species that have been recorded at the sites, but try to emphasise the species that may be of most interest to European birdwatchers. A complete list of all the birds seen in Morocco is included at the end of the guide.

The maps of the larger areas only show the main roads and tracks. To obtain more information on these areas, buy the large-scale maps described on page 17. Bear in mind that the condition of the roads and tracks can change quickly from one year to another and sometimes from one hour to another!

The Straits of Gibraltar
(1a – Tangier Peninsula)

This is one of the best places in the world to observe migration. Only 15 kilometres separate Africa from Europe, and the spectacular movement of migrants across the Straits should not be missed. More than 250 species, mainly European, have been noted crossing the Straits. Most follow a NE-SW axis and some are much more conspicuous than others. The raptors are the most spectacular, but the migration of shearwaters, storks, flamingos, gulls, bee-eaters, larks, swallows, wagtails and finches can also be quite impressive.

Location

The close proximity of Tangier makes it the ideal base for exploring the Straits. As one of Morocco's busiest ports it is very well served by sea, air, railway and road. If coming from Spain, the Straits can be crossed to Ceuta (the shortest crossing) and Morocco entered from there.

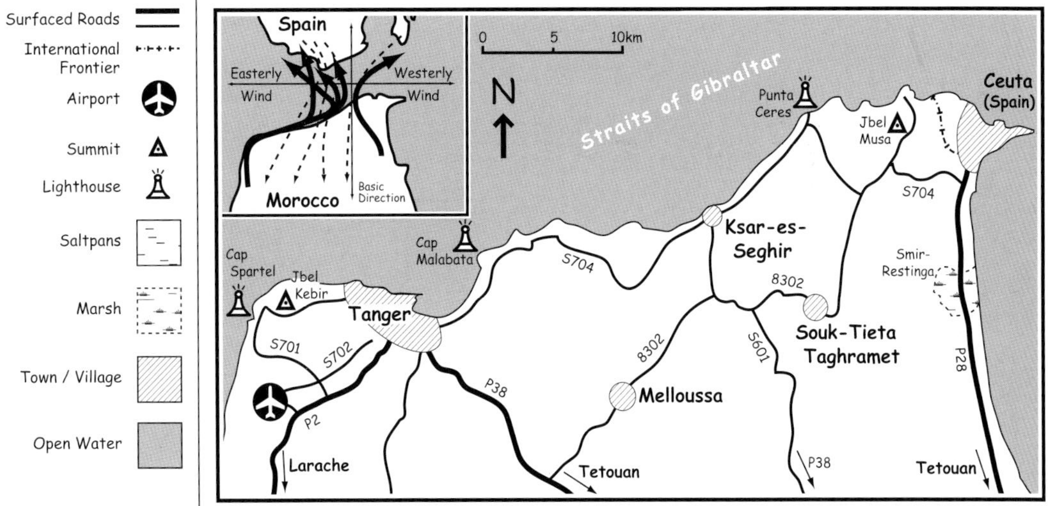

Accommodation

It is not difficult to find accommodation in Tangier. The list of recommended places below is far from comprehensive but includes a selection to suit all tastes. The least expensive, basic hotels are situated in the medina, near the busy and noisy Petit Socco square. The best choice in the medina is the more expensive 'Continental' (❷ 36 rue Dar el Baroud. ☎ 039.93.10.24, Fax 039.93.11.43). Moderate hotels in quieter areas include the 'Marco Polo' (❷ / ❸ corner rue El Antaki – Avenue d'Espagne. ☎ 039.94.11.24, Fax 039.94.22.76), the 'Bristol' (❷ 14 rue El Antaki. ☎ 039.93.43.47 or 94.29.14) and the 'Miramar' (❷ 168 Avenue des FAR. ☎ 039.94.17.15, Fax 039.94.36.28), all three are close to the sea front. The more expensive hotels include the Hotel 'Tarik' (❷ route de Malabata. ☎ 039.34.09.49) and the 'Rembrandt' (❶ corner Boulevard Mohamed V and Avenue Pasteur. ☎ 039.33.33.14, Fax 039.93.04.43). The 'Miramonte' camping site is at the west end of the city, the 'Achakar Grottes d'Hercule' site at Cap Spartel. The city of Tétouan, in the southeast of the peninsula, is also well served by hotels.

There are many restaurants in Tangier providing a wide variety of foods. Some of the least expensive can be found around Petit Socco square while most of the moderate hotels have their own restaurants.

Strategy

Migrants pass through the Tangier Peninsula during most months of the year and occasionally the last of the autumn migrants and the first of the spring may be seen together (for example storks or swallows). The most important movements occur from March to May and August to October.

Visible migration takes place in the morning and the afternoon. Most activity begins about two hours after sunrise and continues into the afternoon. The important visible passages generally stop between one and three hours before sunset. The route of the migrants varies according to climatic conditions. As a general rule, the movements take place in the east of the peninsula when winds come from the west (mainly in spring and November), and in the west when winds blow from the east (mainly in September-October).

The spring movement of raptors varies according to the wind conditions as follows. With strong winds from the east, the migrants are noted on the western coast of the peninsula, from Larache and particularly near Cap Spartel, where their densities increase with the wind. They then follow the southern side of the Straits eastwards and leave Africa between Ksar-es-Seghir and Jbel Moussa. The stronger the wind, the lower they fly. With westerly winds, they travel mainly along the east of the peninsula, following the southern side of the Straits westwards and leaving at Punta Ceres (see map insert, page 23). With gentle winds, birds leave from the entire eastern third of the peninsula and when the wind is very weak, birds leave from everywhere, often at very high altitude. Hundreds of raptors such as kites, harriers and falcons, gather to roost in the scrub of Jbel Kebir or in the forest at the west of Tangier.

In autumn, with westerly winds, most raptors reach Morocco near Punta Ceres (the favoured arrival point, being closest to Spain and creating good thermals). On the other hand, easterly winds spread them west of Punta Ceres and they can even miss the southern side of the Straits altogether.

Birds

The table shows the migration periods of the commonest raptors which cross the Straits and the estimate of their numbers in autumn (after Pineau and Giraud-Audine 1979). Almost all the species follow the general pattern that has been described, but there are some exceptions. Harriers, Accipiters, Ospreys and falcons are less concentrated by the Straits and move on a broader front. Short-toed Eagles are rarely noted at Cap Spartel, Booted Eagles are rare in the west of the peninsula, and Lesser Kestrels are mainly seen in spring. Storks behave in the same way as raptors.

Large numbers of shearwaters move through the Straits. Most Cory's are noted from mid-February to early May (eastward passage and in October-November (westward passage), though small numbers can be sighted in other periods. Balearic Shearwaters are present throughout the year, with a marked post-breeding movement westward from the end of May through August (peaking in late June) and a more prolonged pre-breeding movement eastward from October through to March. Yelkouan Shearwaters are mainly seen between late June and September; the Straits lie at the western edge of the species' post-breeding distribution and are an important feeding and moulting area. The Sooty Shearwater is a scarce passage migrant and winter visitor. Little Shearwater is an accidental visitor (3 sightings, all in August).

The passage of European Bee-eaters, swifts and passerines is light, except in spring when there are strong easterly winds, and they are frequently observed in huge numbers near Cap Spartel.

Several eastern migrants have been seen, far from their normal routes, mainly in spring. These include Pallid Harrier, Spotted Eagle, Red-footed Falcon, Great Snipe, Marsh Sandpiper, White-winged Black Tern, Richard's Pipit, Red-breasted Flycatcher, and Red-backed Shrike. Other rarities for Morocco have included Sabine's Gull, Guillemot, Black Vulture, Purple Sandpiper, and Iceland Gull.

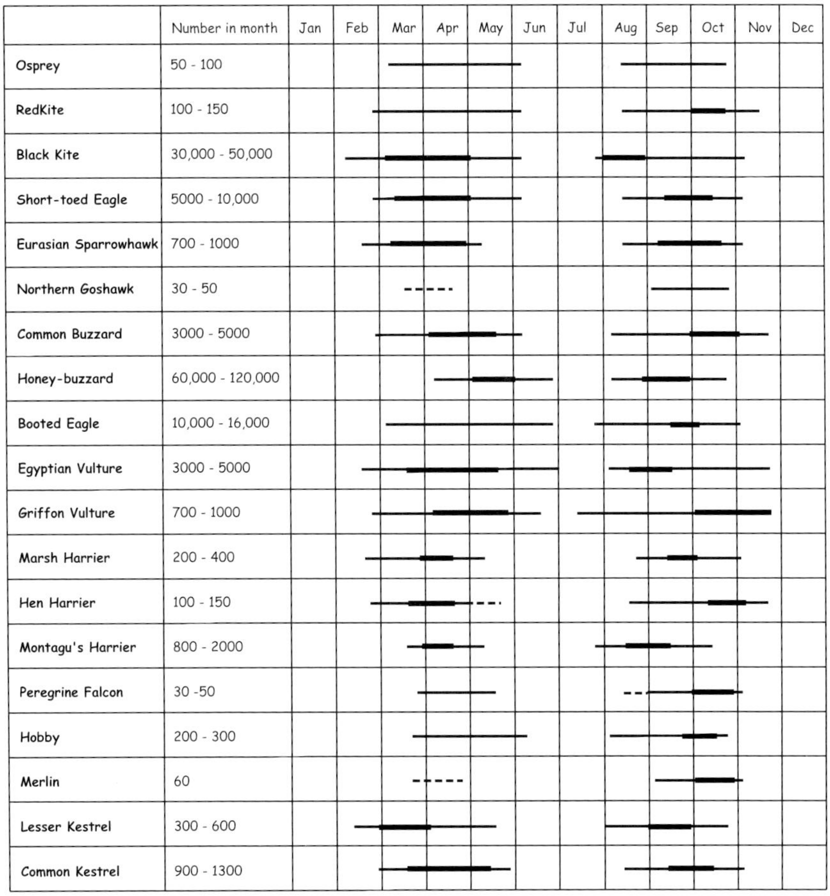

Other Wildlife

The Straits are an exceptional area to observe cetaceans especially from boats. The commonest are Bottlenose and Common Dolphins, and Long-finned Pilot Whale.

Asilah and the mouth of the Oued Tahadart

(1a – Tangier Peninsula)

The region of Asilah is one of the best sites in Morocco for Great Bustards and wintering Common Cranes. To the north of Asilah, the mouth of the Oued Tahadart (also called Oued Mharhar, or Oued Hachef) and the nearby mudflats and marshes are excellent for a wide variety of water birds and have generated some Moroccan rarity records.

Location

Asilah lies along the Atlantic coast on the main road (P2) between Tangier and Larache. The most interesting spot for the bustards and the cranes is some 11km to the northeast of the town, due east of Briech (Briex on the Michelin map).

The Tahadart mouth is 13km to the north of Asilah, west of the 'Pont Mohammed V' on the Michelin map.

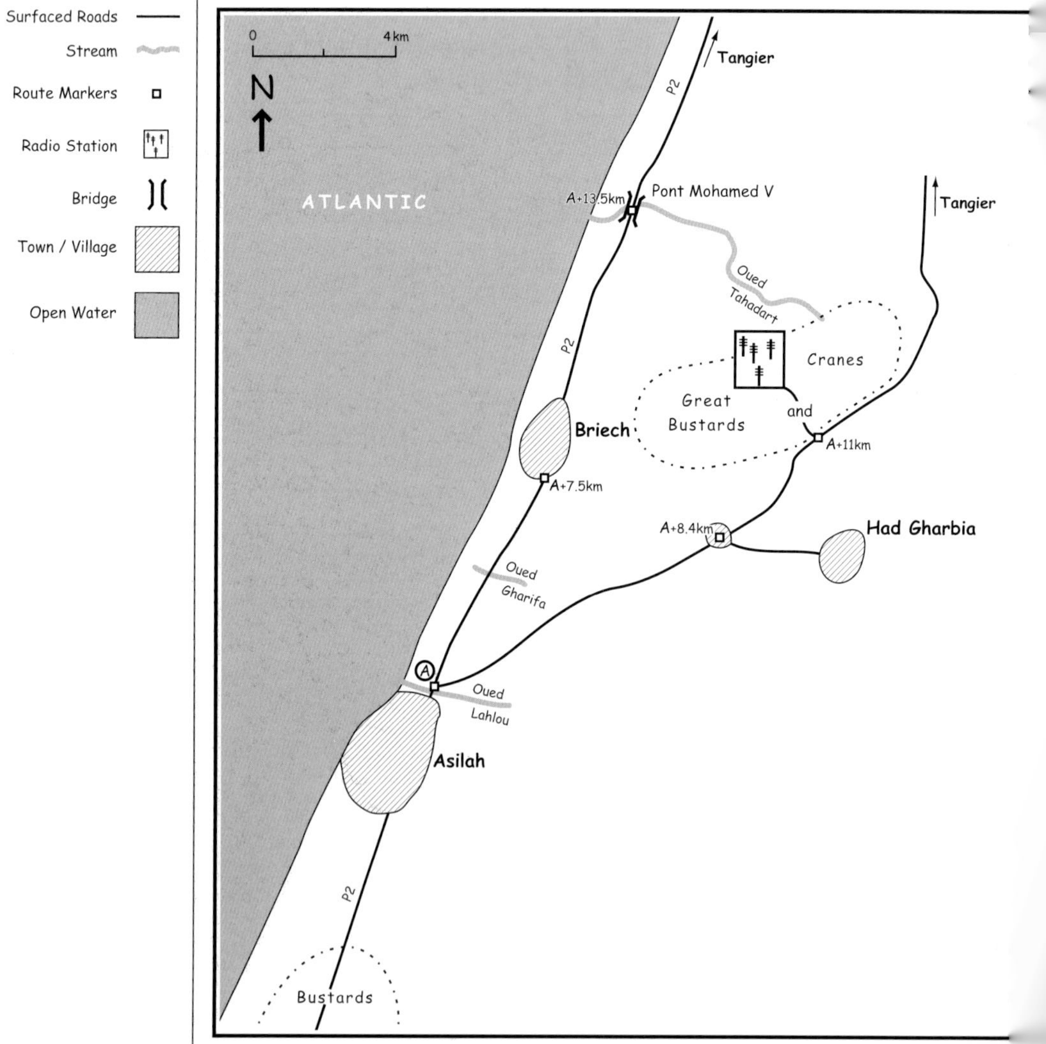

Accommodation

Asilah is a nice little town and has become a popular Atlantic coast summer resort. It is well equipped with hotels and in 2002 our favourites included the 'Zelis' (❶ 10 Avenue Mansour Eddahbi. ☎ 039.41.70.69, Fax 039.41.70.98), 'Oued El Makhazine' (❷ Avenue Melilla. ☎ 039.41.70.90, Fax 039.41.75.00) and 'Mansour' (❸ 49 Avenue Mohammed V. ☎ 039.41.73.37, Fax 039.41.75.33). Outside the town, the 'Al Khaima' (❶ Route de Tanger. ☎ 039.91.74.28, Fax 039.91.75.66) can also be a good choice, especially in summer because of its swimming pool. In any case, it is wise to call in advance as some of these hotels can be closed outside the tourist season.

A few campsites are also available, including the 'As-Saada' (☎ 039.91.73.17) and the fallback choice 'Echrigui' (☎ 039.91.71.82), both on the road to Tangier.

For those intending to go birding at the Lower Loukkos marshes as well, an alternative solution is to stay in Larache (see page 28).

In the tourist season, dozens of restaurants – mostly installed along the walls of the old city – serve fresh fish. At other periods, the choice is more restricted and in October 2000 we found the best value for money at the 'Séville' (18 Avenue Imam Asili. ☎ 039.41.85.05, mobile 063.07.34.84).

Strategy

The town itself is not very good for birds except for Spotless Starlings, but the coast can be rewarding for birds migrating over the sea. Nevertheless most birders concentrate their efforts on Great Bustards and Common Cranes.

The best place for these birds is the fallow lands and rounded hills to the east of the village of Briech. A good vantage point is on the P2 near Briech (7.5km from the Oued Lahlou, on the northern edge of Asilah); stop on the roadside and scan to the east. Another good area is near the Had Gharbia radiocomm station: just after the bridge over the Oued Lahlou, on the northern edge of Asilah, take the road to the northeast signposted 'Had Gharbia 11 – Barrage du 9 Avril 1947 25'. After 11km, a 'no-entry' road leads to the radiocomm station; stop nearby and watch carefully in the fallows for the bustards (some birders climb on top of the hill mid-way between the 'no-entry' sign and the station gate to gain a vantage point but we do not recommend doing so as we encountered major problems with the security personnel here in October 2000; this is a sensitive, military area).

Another good yet less famous area for bustards is some 6.5km to the south of Asilah (from the Stadium), on the hills and fallows on both sides of the P2. Up to 40 Little Bustard territories and one male and 22 Great Bustard females were counted here in the early 2000s. Unfortunately, the Asilah-Tangier highway is under construction, and the site might not be as good in future years.

The mouth of the Tahadart, 13.5km north of Asilah on the P2, is worth a visit. Stop just before the Pont Mohammed V for excellent views of the dunes that fringe the mouth (good for gulls), and of the salt marshes upstream from the bridge (looking east) which are good for ducks and waders.

Birds

The Moroccan population of Great Bustards was estimated as 90–133 birds in the late 1990s, and most of them were concentrated in the areas described here. Females were three times more numerous than males and this indicates that this small, endangered population is unbalanced; it suffers from loss of habitat, hunting and poaching, and collision with overhead power lines. Avoiding any disturbance to these birds is thus essential so please do not wander in their territories; stay on the roadsides or on the tracks but do not give up too quickly as they are often difficult to see when they forage and hide in the furrows.

Briech is also famous for wintering Common Cranes. The first autumn migrants have been noted in mid-October and the last spring migrants in early March but numbers peak from November through to February; up to 600 birds have been recorded.

The salt marshes of the Oued Tahadart attract a wide range of ducks and waders both in migration periods and in winter. Among the most conspicuous are Greater Flamingos (up to 246 in January 1990) and Spoonbills which are regular at the site and attempted to breed in 1967; this is one of the last breeding places for Marsh Owl.

The mouth of the Oued often acts as a resting place for gulls and a rarity is always possible among the numerous Yellow-legged and Lesser Black-backed Gulls: on 18 October 2000, we saw four Ring-billed and one Audouin's Gulls and four Royal, one Caspian, and two Sandwich Terns, while a *calidus* Peregrine was sitting on top of the dunes.

Other rare species here have included Pallid Harrier, Spanish Imperial Eagle, Sociable Lapwing, Marsh Sandpiper, Sabine's Gull, and Roseate and Arctic Terns.

The cork-oak woods near Asilah are the habitat of lots of interesting species including several raptors (Black-shouldered Kite, Booted Eagle, Hobby and perhaps Goshawk), European Nightjar, Levaillant's Green Woodpecker (rare), Great Spotted Woodpecker, Woodlark, Dartford Warbler, Golden Oriole, Woodchat Shrike, and lots of finches including the superb Moroccan Greenfinch (*vanmarli*). Woodcock is a fairly common winter visitor. The nearby open habitats are home for Barbary Partridge, Calandra Lark, and Corn Bunting.

Larache and the Lower Loukkos marshes (2a – Rharb)

The salt pans near Larache and the marshes of the Lower Oued Loukkos are among the most interesting wetlands in northern Morocco. They are home to such species as Purple Swamphen, Red-knobbed Coot, and Moustached Warbler, and form a superb stop on the migration route for numerous water birds.

Location

Larache city lies some 90km south of Tangier, on the Atlantic coast. The salt pans are just downstream from the bridge where the main road from Tangier to Rabat (P2) crosses the Oued Loukkos, while the marshes (also called Aïn Chouk or Boucharen marshes) are upstream from the bridge.

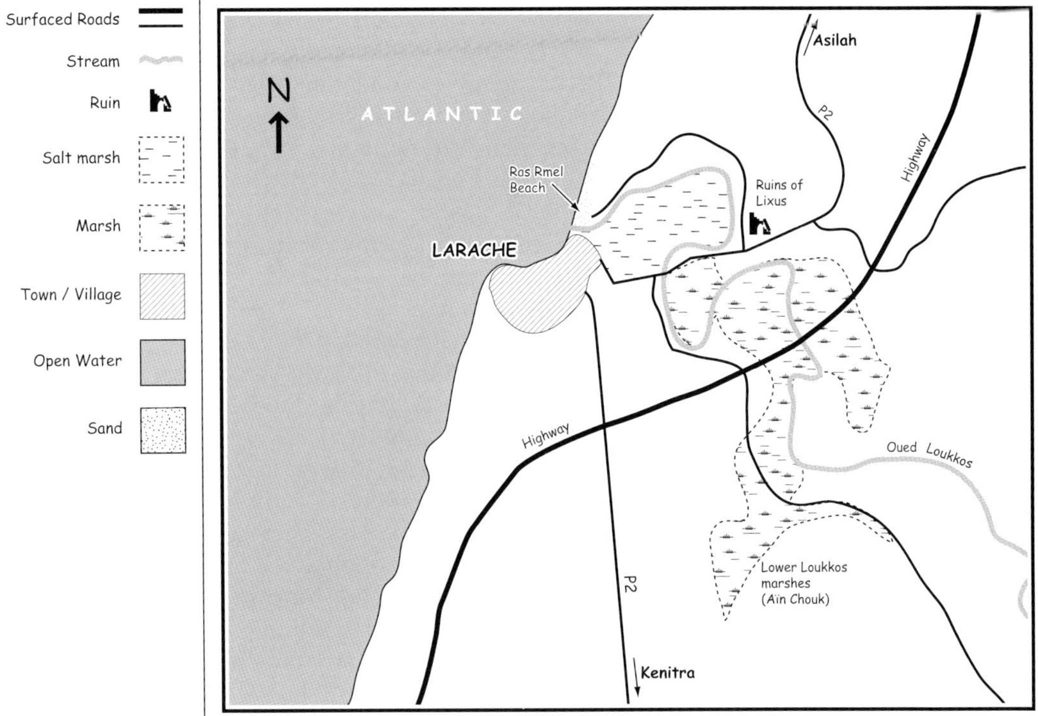

Accommodation

In 2002, there were only a few decent hotels at Larache. The best choices included the 'Riad' (❶ 87 rue Mohamed Ben Abdallah. ☎ 039.91.26.26, Fax 039.91.26.29), the 'España' (❷ 2 avenue Hassan II. ☎ 039.91.31.95), and the 'Pension Malaga' (❸ Passaje del Teatro. ☎ 039.91.18.68).

For those intending to go looking for the bustards near Asilah a better solution is to stay in a hotel there (see page 26).

Strategy

The migration periods of March-April and September-November are probably the best times, as the variety and quantity of birds are greatest then, but winter is also rewarding.

It is well worth spending half a day or more exploring the two main sites. Try to stay near the marshes at sundown – they can be superb!

To get to the marshes from Larache, follow the P2 to the north and, just before the bridge over the Loukkos, turn right on a tarmac road signposted 'Barrage de garde sur Oued Loukkos'. A first excellent marsh is at 3.5km, the main marsh is on the right at 5-6km. A bit further, near the barrage on the Loukkos, the grasslands are sometimes good for Little Bustards in winter.

These semi-permanent, fresh water marshes (some 1350 ha) are partly state- and partly privately-owned. Hunting is normally forbidden, but they suffer from various disturbances including poaching, agriculture encroachment, grazing, and pesticide and fertilizer spreading. They deserve good protection.

The salt pans downstream from bridge over the Oued Loukkos on the P2 are hard to miss. Stop on the wide roadsides: the birds are only a few metres away. Be careful of fast driving cars and trucks.

If time allows, follow the right bank of the Oued Loukkos down to its mouth; a tarmac road (signposted 'Plage de Ras Rmel') starts near the Roman ruins of Lixus and gives good views over the nearby mudflats; the pine woods on the dunes attract hundreds of finches, and Spotless Starlings breed in the man-made constructions. At the mouth, several small restaurants serve fresh fish.

Birds

The marshes are excellent for a wide range of ducks. They have been recognized as a Moroccan nationally important site for Greylag Goose (mean and max numbers from 1990-1994: 39/96), Wigeon (1019/1948), Gadwall (294/870), Common Teal (300/973), Mallard (201/284), Pintail (178/760) and Northern Shoveler (1379/1882), but also for Ferruginous Duck (30/68), and for Red-crested Pochard (60/232), whose numbers have increased in recent years (up to 410 in December 2000). Marbled Teal sometime occurs (e.g. 24 on 13-14 September 2001).

They are also good for large waders, including Glossy Ibis (up to 89 in December 2000), Little Egret, Grey and Purple Herons, White and Black Storks, Spoonbill, Greater Flamingo, and Squacco Heron. Nevertheless, the most sought-after birds are probably Red-knobbed Coot and Purple Swamphen, which are fairly common here (e.g. c.1100 Red-knobbed Coots early March 2002; scan the reedbeds fringe for the latter). A wide range of small waders will be seen, including perhaps some uncommon species such as Marsh Sandpiper or Jack Snipe. Black-winged Stilt is often very common in migration periods, and also breeds here.

Numbers of wintering Little Bustards have decreased over recent decades (a thousand birds in January 1964, a hundred in January 1990), but a few dozen should still be seen.

Various warblers breed here, including Cetti's, Savi's, Moustached, Fan-tailed, Reed, and Great Reed. The local race of Reed Bunting (*witherbyi*) is now very rare. Rare breeders also include Little Bittern and Baillon's Crake.

Species rare or accidental in Morocco have included White-fronted Goose, Pectoral Sandpiper, Citrine Wagtail, Aquatic Warbler, Penduline Tit, Carrion Crow, and Tree Sparrow. The elusive Andalusian Hemipode has been recorded near Larache.

The salt pans downstream from the bridge on the P2 are excellent for ducks, waders, gulls, and terns. Besides the commoner species, some less common or even rare species for Morocco can be found: Mediterranean, Slender-billed, Ring-billed, Herring, Iceland, and Glaucous Gulls, and Caspian, Royal, Roseate, and White-winged Black Terns have all been recorded here.

There are few seawatching reports for the Larache area, but it seems promising. More than 1,400 Leach's Storm-petrels, 35 Grey Phalaropes, a few European Storm-petrels and three species of skua were sighted in a couple of hours seawatching at Christmas 1996.

The cork-oak woods near Larache are the habitat of lots of interesting species including several raptors (Black-shouldered Kite, Booted Eagle, Hobby and perhaps Goshawk), European Nightjar, Levaillant's Green Woodpecker (rare), Great Spotted Woodpecker, Woodlark, Dartford

Warbler, Golden Oriole, Woodchat Shrike and lots of finches including the superb Moroccan Greenfinch (*vanmarli*). Woodcock is a fairly common winter visitor.

The first records of Collared Dove and Jackdaw at Larache date from only 1990; in town, numerous Pallid Swift and smaller numbers of Little Swift should be seen.

The Merjas of the Rharb – Halloufa, Bargha and Oulad Sgher

(2a – Rharb)

Apart from Merja Zerga (see page 34), these are the three main merjas (lagoons and marshes) of the Rharb. They attract impressive numbers of birds in migration periods and in winter.

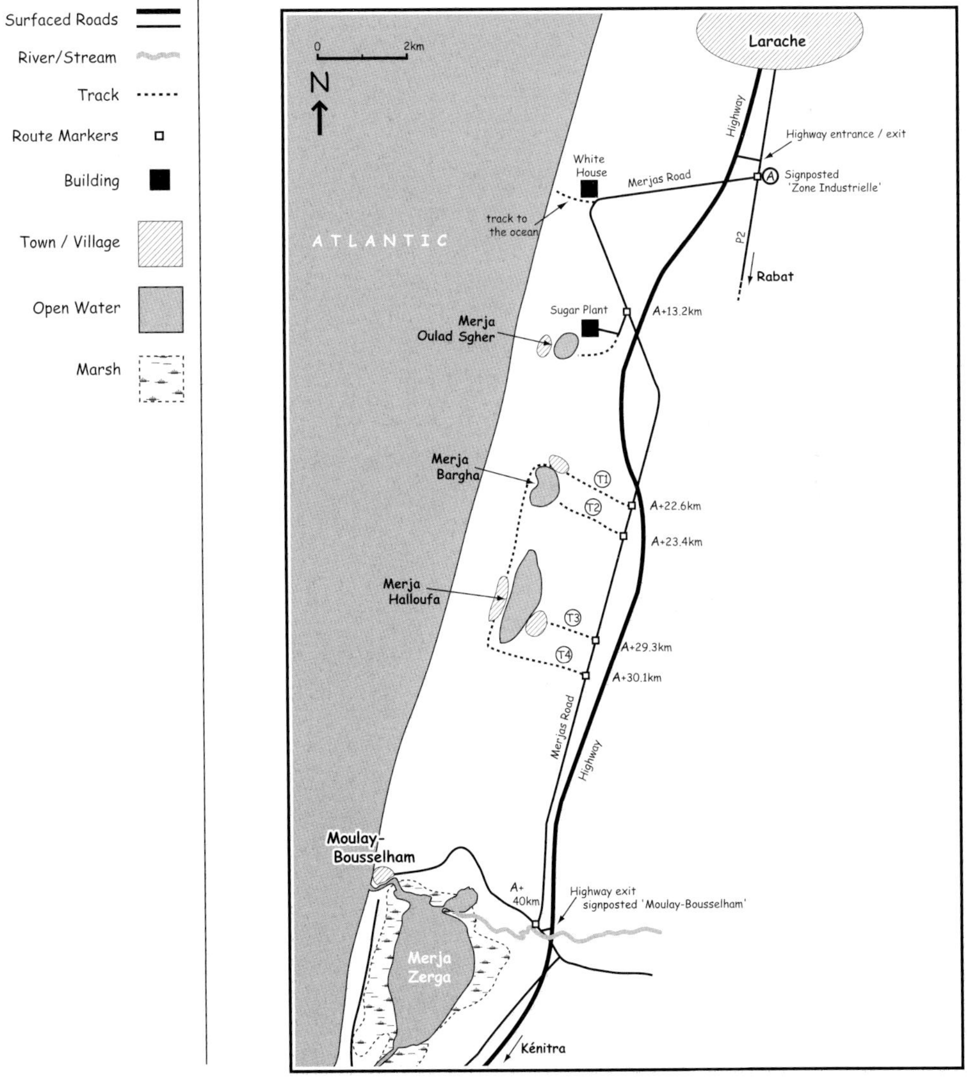

Location

These three lagoons are spread along the Atlantic coast between Larache and Merja Zerga and are separated from the ocean by a row of consolidated dunes. From north to south they are Oulad Sgher, Bargha, and Halloufa.

Access to the lakes is via the 'merjas road' that can be taken either coming from the north (Larache) or from the south (Kénitra). The sketch shows the location of the merjas in relation to the junction of the 'merjas road' with the main P2 road, just to the south of Larache (point A). Distances are given from this point, and those with cars should set the odometer to zero at this point. Some of the published maps (including the Michelin) are not accurate and it is better to refer to the sketch in this guide even though the distances have been recorded using the odometer of a rented car and may be slightly out.

Coming from the north (Larache) on the main highway, leave the highway at the 'Larache sud' exit and follow the P2 to the south for some 900 metres. Then turn right on to the 'merjas road', which is signposted 'Zone industrielle'. Coming from the south (Kénitra), leave the highway at the 'Moulay Bousselham' exit then turn to the right and follow signs to 'Larache-Chouarfa'; this is the other end of the 'merjas road'.

To reach Oulad Sgher coming from Larache, at A+13.2km take the little tarmac road that leads to a sugar plant just before the 'merjas road' passes under the highway. Follow this tarmac road for 1.8km then carry on straight down the track; Oulad Sgher in on the far right after crossing an irrigation channel.

Merjas Bargha and Halloufa are visible from the 'merjas road'. Just after the 'merjas road' passes under the highway again two tracks at A+22.6km and A+23.4km (T1 and T2) lead to the eastern bank of Merja Bargha. Two other tracks at A+29.3km and 30.1km (T3 and T4) provide access to Halloufa. The T3 track (skirted by a row of telephone poles) leads to a village; turn right in the village between rows of prickly pears to go closer to the shore. The T4 track (skirted by a row of concrete poles) leads to the southern shore of Halloufa, then skirts its western edge and goes to Bargha through some villages (coming back from Bargha to the 'merjas road' through Bargha village is difficult and we recommend taking the T4 back).

Be careful not to venture too close to the merjas with a car due to the risk of becoming bogged down.

Accommodation

No accommodation or food is available near the merjas; the closest hotels are at Moulay Bousselham (page 34) and Larache (page 28) and it is wise to take a picnic.

Strategy

A whole day is needed to explore the three merjas as birds are generally plentiful, especially during the migration periods (September-November and March-May) and in winter.

For the best light conditions, visit Oulad Sgher and the eastern side of Bargha and Halloufa in the morning and the western banks in the afternoon. The size of the merjas varies according the quantity of rain. A telescope is essential here.

Oulad Sgher is the least known of the three merjas perhaps because of its remote location.

Birds

In winter some Black-necked Grebes should be present among the numerous Little Grebes (up to 1,950 on 30 October 1991 at Bargha). Great Crested Grebes breed (c.50 pairs at Bargha in 1999). The merjas are also good for large waterbirds and common species include Little and Cattle Egrets, Grey Heron, Greater Flamingo, and Spoonbill. Glossy Ibis, Little Bittern, Night and Squacco Herons are less common though up to 10+ Squacco and 20 Night Herons were at Bargha on 20 September 1999; Great Egret is an accidental visitor.

The merjas are excellent for ducks. Mallard, Pintail, Gadwall, Common Teal, Garganey, Northern Shoveler, Pochard and Tufted Duck are among the commoner species, while Marbled and Ferruginous Ducks are rarer. Red-crested Pochard is often common in winter at Halloufa (up to 300 in January 1994) and bred at Bargha in 1995.

The White-headed Duck disappeared from Morocco in the early part of the 20th century and has only been an accidental visitor since then. But the number of records has increased quite recently (after the increase in the Spanish population) and up to eight birds have been sighted at Bargha. Unfortunately, Ruddy Duck has also appeared (40+ records at Bargha, with a maximum of 11 in January 1994, and 25+ records at Halloufa, maximum of eight in January 1995) and hybrids have been recorded although breeding has not been definitely proven. There were for example six hybrid males and one hybrid female at Bargha on 22 December 1996.

Osprey, Marsh Harrier, and Kestrel are among the commoner raptors but Peregrine, Lesser Kestrel, Sparrowhawk, and Booted Eagle have also been recorded. Some pairs of Marsh Owl still breed in the area.

Most of the wader species have been recorded at these sites. Black-winged Stilt and Northern Lapwing breed at Bargha (150-200 pairs in 1998 for the former, 20-30 pairs for the later); the last record of Slender-billed Curlew at Bargha dates from November 1991 (five birds). At Bargha also, some Red-knobbed Coots (maximum 400 on 20 October 1997) may be seen among the commoner Eurasian Coots (maximum 5,500 in January 1992), and some Purple Swamphens may be spotted in the marshy vegetation.

The three merjas often attract large flocks of gulls (for example 12,000 Lesser Black-backed at Oulad Sgher in January 1999) which can also include more unusual species such as Slender-billed Gull, or Caspian, Whiskered, Black and White-winged Black Terns.

The coastal lawns, hedges and farms attract species such as Stone-curlew (up to 300 between Bargha and Halloufa on 16 January 1994), Crested Lark, Stonechat, Fan-tailed and Sardinian Warblers, and Spotless Starling. The first Jackdaw was recorded in the early 1990s, and the species is now fairly common.

Moroccan rarities have included Great Egret, American Wigeon, Ring-necked Duck, Ring-billed Gull, and Penduline Tit.

Moulay Bousselham Reserve: The Merja Zerga

(2a – Rharb)

This internationally important wetland was declared a biological reserve in 1978 and a Ramsar site in 1980. Covering 90km², it includes one of the largest lagoons in Morocco (the Merja Zerga, 30km²).

The lagoon, surrounded by a wide belt of low vegetation, is an important staging and wintering place for ducks and waders, and a breeding site for some uncommon birds. It also acts as a sanctuary when climatic conditions are unfavourable in Europe, especially in southern Spain.

Location

This wetland lies on the Atlantic coast, 40km south of Larache and 70km north of Kénitra. The quickest way to get there by car is to drive along the Larache-Kénitra highway, and exit at 'Moulay Bousselham'. A more pleasant route which can be taken from the south follows the S206 from Kénitra for 25km and then the 2301 that runs along the Atlantic coast.

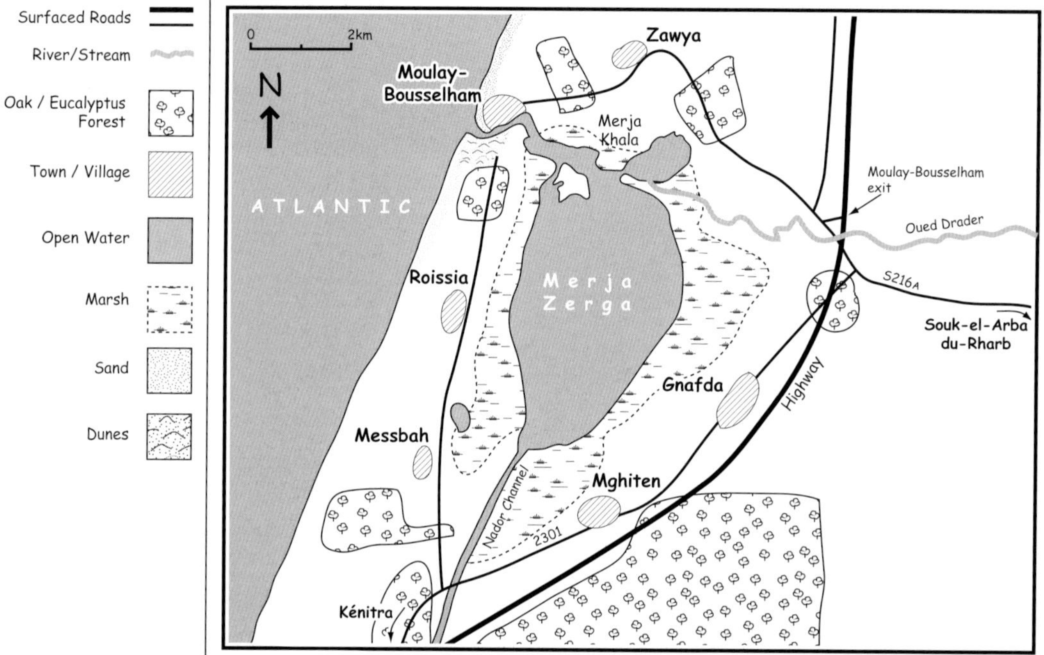

Accommodation

The village of Moulay Bousselham has a hotel which is currently (2002) in poor condition ('Le Lagon' ❷ ☎ 037.43.26.50), a camp site 500 metres before entering the village, and some restaurants nearby (the 'Café Milano' keeps a bird-log). As the site can be reached quickly by road, we prefer using accommodation in one of the nearby cities such as Larache (see page 28). Those camping may like to buy fresh fish, which are available when the local fishermen return in the afternoon.

Strategy

Because of the wealth of birdlife at the site, it is worth spending at least a day here.

Winter is the season when birds are most numerous and it is possible to see more than 1,000 Greater Flamingos, 30,000 ducks, 40,000 Eurasian

Coots (up to 60,000 in January 1981), and 50,000 waders. Visits to the site can also be rewarding in spring and autumn, when a lot of passage migrants stop here and breeding birds are present. Try to avoid the middle of hot days because of the heat haze. Rain occurs regularly from late autumn to early spring and the meadows are then often very wet. It is essential to have rainwear and wellingtons if intending to walk around the site at that time of year.

Visitors with a car should travel around the lagoon on the 2301: the road gives good general views but a telescope is essential as most birds are usually fairly far away. Ducks are often best viewed from Moulay Bousselham and from the eastern shore (also good for small waders). During the breeding season White Storks can be seen at close range attending their nests, built on the low huts of the villages. At the south of the lagoon, where the road crosses the Nador Channel, there is a new road that hugs the western bank. This road leads to the southern edge of the narrow channel, just in front of Moulay Bousselham and provides other interesting views of the Merja. For visitors without a car, the Merja Kahla (also called Merja Mellah) to the northeast of the Zerga is easily reached by walking from Moulay Bousselham and is a good spot to observe most species.

At Moulay Bousselham, consider asking a fisherman to take you for a boat ride on the lagoon. Seawatching can also be rewarding with Moroccan rarities such as Madeiran Storm-petrel and Fulmar recorded, as well as commoner species such as Balearic Shearwater, skuas, gulls and terns.

Birds

Half of all the ducks and the small waders wintering in Morocco north of latitude 30°N are found at the Merja Zerga. Among the ducks, Wigeon is the commonest (c.20,000, 80% of the population wintering in the country), followed by Common Teal (c.7,000), Mallard and Northern Shoveler (c.2,000 each), Pintail, Shelduck, and Gadwall. Marbled Duck is irregular here, with a maximum count of 800 in January 1964, but is generally much rarer, with less than 25 birds normally present. Numbers of Red-crested Pochard have increased in recent years in Morocco, and some birds are occasionally seen here. The rarest of the regular waterfowl are probably Ruddy Shelduck and Ferruginous Duck.

Slender-billed Curlew

Of the waders, the most abundant are Dunlin (c.20,000), Black-tailed Godwit (c.10,000), Ringed Plover (c.5,000), Redshank (c.3,000), Avocet and Kentish Plover (c.2,000 each), and Common Snipe.

In the 1980s and 1990s very small numbers of wintering Slender-billed Curlews were found at this site, but none have been seen since 1998. It may still be worth checking their favourite place in the muddy freshwater meadows or on fallow fields to the east of the merja, in case one reappears some day (the north of the village of Gnafda was the best).

Little Egrets, Grey Herons, White Storks and geese (mainly Greylag, sometimes Bean, White-fronted or Brent) are also present.

1,500-2,500 Greater Flamingos and 10-40 Spoonbills winter here; their numbers increase during the migration periods in September-October and especially in March-April; in April 1982 up to 6,700 Greater Flamingos and 115 Spoonbills were counted.

The total number of birds present at the lagoon is smaller in spring and autumn, but there is a greater variety. From early March to mid-May and from late August to late November, the majority of the Western Palearctic migrant waders and a lot of migrant passerines are likely to be seen.

During the breeding season (March-June), the area around the lagoon supports a good number of species. Turtle Doves, Crested Larks, Yellow Wagtails, Zitting Cisticolas and Corn Buntings are abundant. Black Kites, Marsh and Montagu's Harriers, and Ospreys (1-5 non-breeding birds present throughout the year) should be seen. At the south end of the lagoon 200-500 pairs of Collared Pratincoles nest as well as the most southern breeding colony of Northern Lapwing (20-30 pairs). This is one of the best sites to see Marsh Owl and Red-necked Nightjar, and both are best looked for after sunset. The owl can be seen around the lagoon and especially at the Moulay Bousselham campsite whose trees often act as a daytime roosting place (20+ birds sometimes noted there in winter); other good places include the northern shore near the Merja Kahla and the southwestern shore near the village of Massbah. The nightjar can be found in the cork oak forest nearby. Other interesting breeding species in the vicinity include Black-shouldered Kite, Little Bustard and Stone-curlew.

Of the gulls and terns, only Little Terns breed (c.25 pairs), but many other species winter or pass through. Mediterranean Gulls have been noted among the commoner wintering species. Spring is the season when the majority of the other species have been observed: Slender-billed and Audouin's Gulls, and Gull-billed, Royal, Lesser Crested, Whiskered and Black Terns. Caspian Terns can be seen throughout the year.

Some rare species for Morocco or for the Western Palearctic have been noted here: Fulmar, African Darter, Great Egret, Lesser Flamingo, Mute Swan, Ring-necked Duck, Pallid Harrier, Allen's Gallinule, Marsh Sandpiper, Ring-billed Gull, Red-throated and Richard's Pipits, and Penduline Tit.

Other wildlife

Several species of amphibians and reptiles can be observed around the lagoon among which are Sharp-ribbed Newt, Mauritanian Toad, and Montpellier and Viperine Snakes.

Lac (Merja) de Sidi Bourhaba
(2a – Rharb)

This attractive lagoon, surrounded by juniper woodland, runs parallel to the Atlantic coast between two rows of consolidated dunes. Its southern half, a biological reserve since 1976 with prohibited access, is part open water and part marsh, and its banks are densely wooded. The northern half is freely accessible and forms a lake fringed by a small marsh. It shelters several hundred wintering ducks and is a breeding place for various interesting species.

Location

Sidi Bourhaba is located 30km north of Rabat, west of the P2, near Kénitra and the village of Mehdiya. The best way to get there is by car from Rabat or alternatively take the Kénitra bus and ask the driver to stop at the junction of the S212, the road to Sidi Bourhaba. From the junction you can then hitch-hike to the lagoon. From Kénitra, grand taxis will drive you there easily.

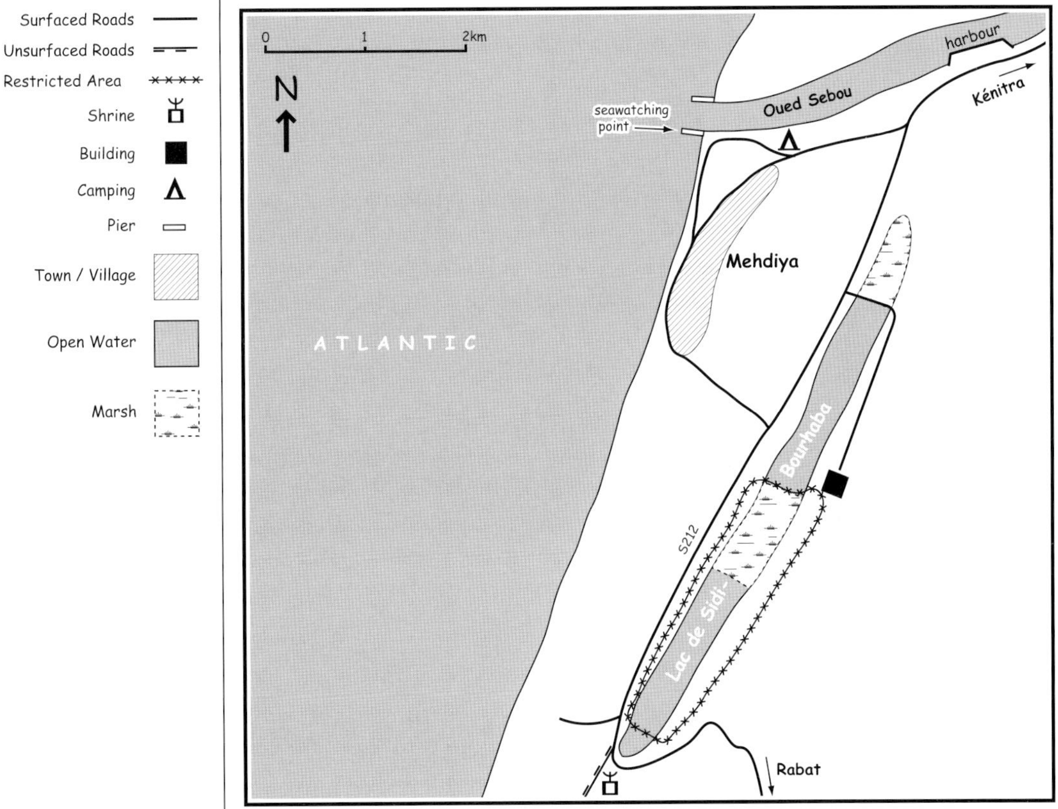

Accommodation

There is no accommodation or shops near the lake, and camping is not recommended. A camping site is open in summer near the Sebou Wadi, and the village of Mehdiya holds only one hotel, the 'Atlantic' (❷ ☎ 037.38.81.16), which was not very good value for money in 2002. The 'Firdaous' (❶ Plage des Nations, Sidi Bouknadel, BP 4008, Salé, ☎ 037.82.21.31, Fax 037.82.21.43), a few kilometres to the south, is located on a nice beach and is probably the best choice if within your price range. Other accommodation is available in Rabat or Kénitra, two large cities with a wide range of hotels. Take a picnic and have lunch near the lagoon; tables and benches have been built under the eucalyptus on the eastern bank.

Strategy

The site can be visited throughout the year, but summer is rather quiet and the birds are less numerous and more difficult to see. Ducks are abundant throughout the winter, and can also be plentiful in September-October, during the autumn migration; a visit in March should produce some migrants and the first breeding birds as well. It is a good idea to spend half a day at the site, but avoid Sundays, when people from Kénitra and Rabat visit in great numbers. Try to stay until sunset, when Marsh Owl may be seen.

The road which overlooks the lagoon provides a good viewpoint and you can walk everywhere, except in the reserve. The western edges are best, especially the small marsh at the northern tip. The rich scrub area between the Merja and the sea is worth investigating for Barbary Partridge, warblers, and Moroccan Magpies. At low tide, it is worth visiting the fishing harbour on the Sebou Wadi for waders, gulls and terns – Purple Sandpiper, Audouin's, Ring-billed and Glaucous Gulls, and Lesser-crested and Royal Terns have been seen here. The tip of the pier at the mouth of the Oued Sebou is a good seawatching place.

Birds

In winter, mixed in with the flocks of European ducks (c.3,000-4,000), Marbled Ducks may be seen (up to 1,700 in January 1982), as well as other rarer species such as Red-crested Pochard or Ferruginous Duck. Waders and herons can be seen along the banks; Cattle Egrets sometimes gather in large night roosts. The woods harbour numerous wintering passerines. At this time, Red-knobbed Coots (c.20 pairs) are harder to distinguish from Eurasian Coots (look for their pale blue bills contrasting with their white frontal shields) but from February or March, as the breeding season starts, their red knobs appear making them easy to spot. At this time Great Crested Grebes (c.10 pairs) and Marsh Harriers begin to display, and at dusk Marsh Owls can be seen hunting over the marsh by the northwest tip of the restricted area and sometimes from the causeway at the north end of the lagoon.

Spotted and Baillon's Crakes have been seen in March. Migration is at its peak in April, and involves such species as Squacco and Purple Herons, Greater Flamingos, Montagu's Harriers, and Avocets. In May and June, sometimes earlier, breeding Black-winged Stilts and families of Marbled Ducks (5-10 pairs in the 1990s) and coots can be seen. A lot of passerines breed around the lagoon and Cetti's, Fan-tailed and Sardinian Warblers are fairly common. A colony of the Moroccan race of Magpie, with a blue patch behind the eye, breeds in the reserve, and Hobbies often come to hunt over the water. Ferruginous Ducks bred in 1997, Red-crested Pochards in 1998 and 2000; Purple Swamphens did so in 1997 and 1998, and are now regular at the site, in small numbers. Black-shouldered Kite and Eleonora's Falcon are not uncommon.

Some rare species for Morocco and the Western Palearctic have been noted at the lagoon including Fulvous Whistling-Duck, American Wigeon, Blue-winged Teal, Ring-necked Duck, and Sabine's Gull.

Seawatching has produced sightings of Cory's, Great, Sooty, Manx and Balearic Shearwaters, European and Leach's Storm-petrels, Pomarine, Arctic, Long-tailed and Great Skuas, Audouin's, Little and Common Gulls, Kittiwake, Guillemot, Razorbill, and Puffin.

Other wildlife

The woodland around the lagoon shelters several reptiles including Common Chameleon and Horseshoe Snake. Spur-thighed Tortoise is common.

The Zaër

(2e – Zaër)

The Zaër form the western part of the Central Moroccan Plateau and contain some of the original cork-oak and Sandarac woods. The rich avifauna (c.100 breeding species) includes some rare birds such as Black-shouldered Kite, Double-spurred Francolin, and Black-crowned Tchagra.

Location

The most interesting part of this region, and the easiest to visit, is located in the Rabat-Casablanca-Rommani triangle. A car is needed to fully explore this area.

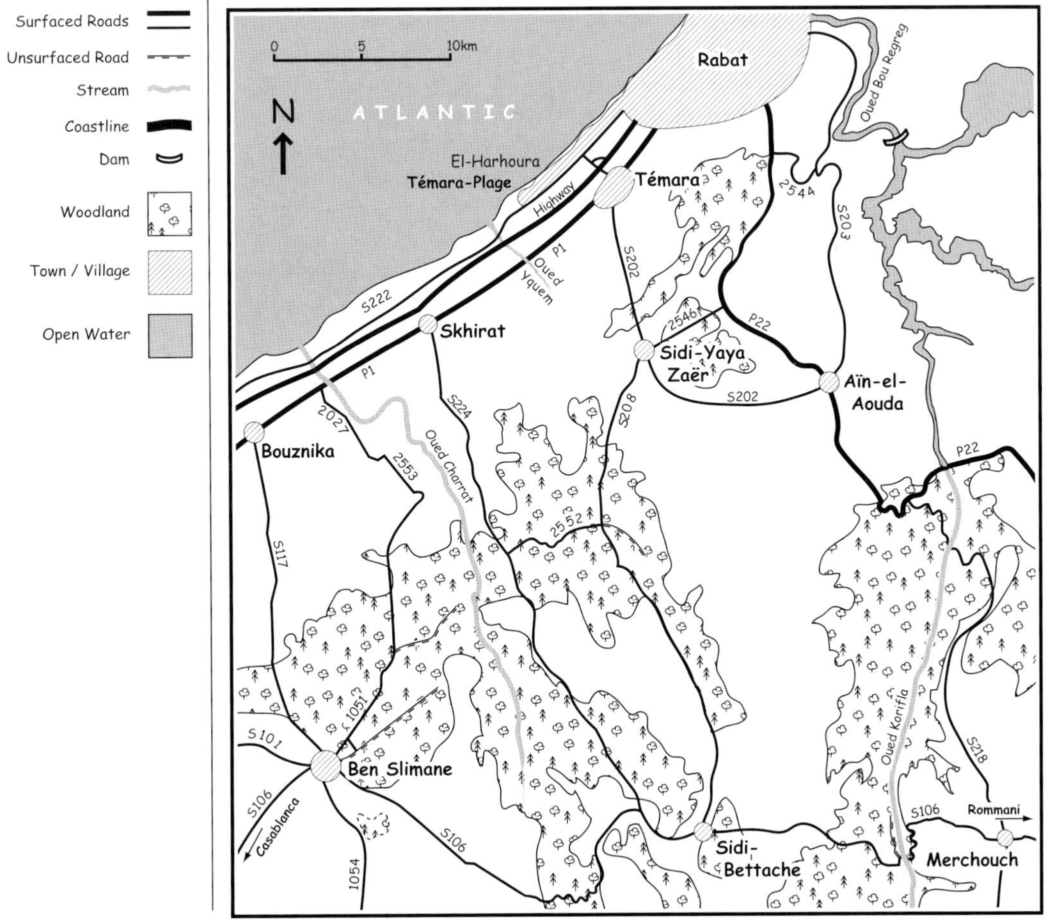

The S202 road to Sidi Yaya Zaër can be hard to find, and the following sketch shows how to get there, coming south from Rabat on the main highway.

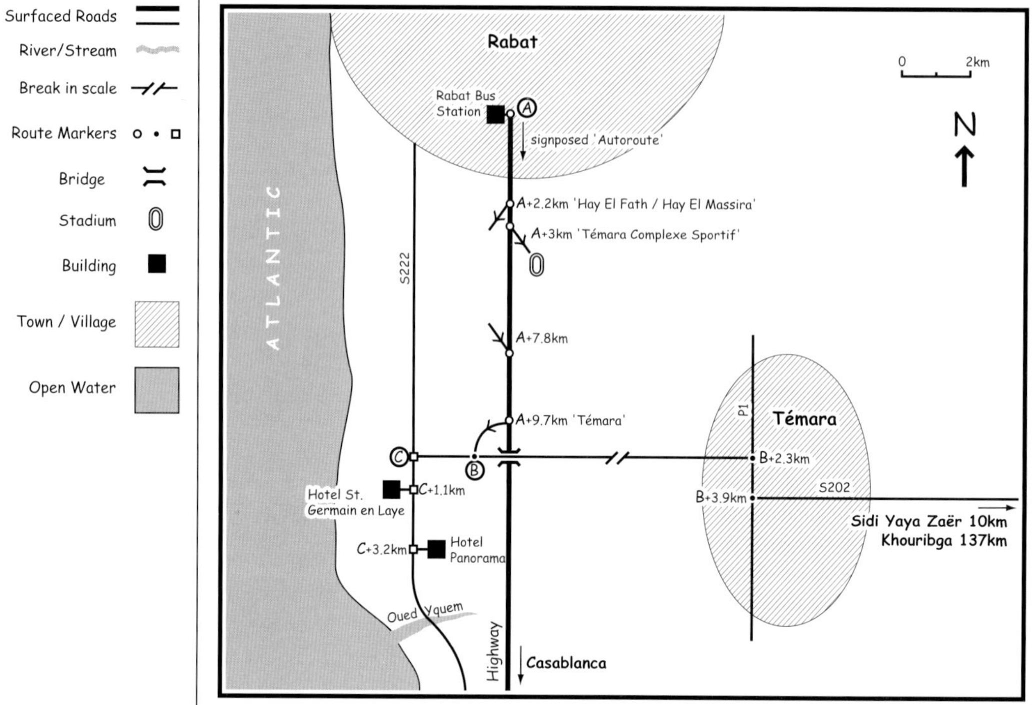

Accommodation

Rabat and Casablanca have numerous hotels, of all types, but those intending to go searching for Double-spurred Francolin early in the morning should stay in the strategically located (though expensive) hotels close to the sea to the south of Rabat, at Témara and Skhirat. These include the 'Saint Germain en Laye' (❶ Témara Plage. ☎ 037.74.42.30 and 48.49, Fax 037.74.48.50), 'La Felouque' (❶ Plage des Sables d'Or. ☎ 037.74.43.88, Fax 037.74.40.69) and 'Panorama des Sables' (❶ Plage des Sables d'Or. ☎ 037.74.42.89, Fax 037.74.48.19). Campsites can be found at Casablanca or Mohammedia.

Strategy

One of the best times to visit the region is early spring (March or April) when migrants should be plentiful. Summer visitors including European Bee-eater, Rufous Bush Robin, and Western Olivaceous Warbler are present from April onwards and leave in September-October. After a rainy winter, there are hundreds of dayas (temporary ponds), an abundance of flowers, and the lush green forests resound with birdsong.

It is wise to start very early in the morning if you wish to look for Double-spurred Francolin near Sidi Yaya; plan to spend at least three or four hours at Sidi Yaya and a full day if you intend to explore some of the numerous tracks which go deep into the forests and down into the valleys. The best area for the Double-spurred Francolin is located in the Royal Hunting Reserve some 14-18km from Sidi Yaya along the road to Sidi Bettache (see sketch).

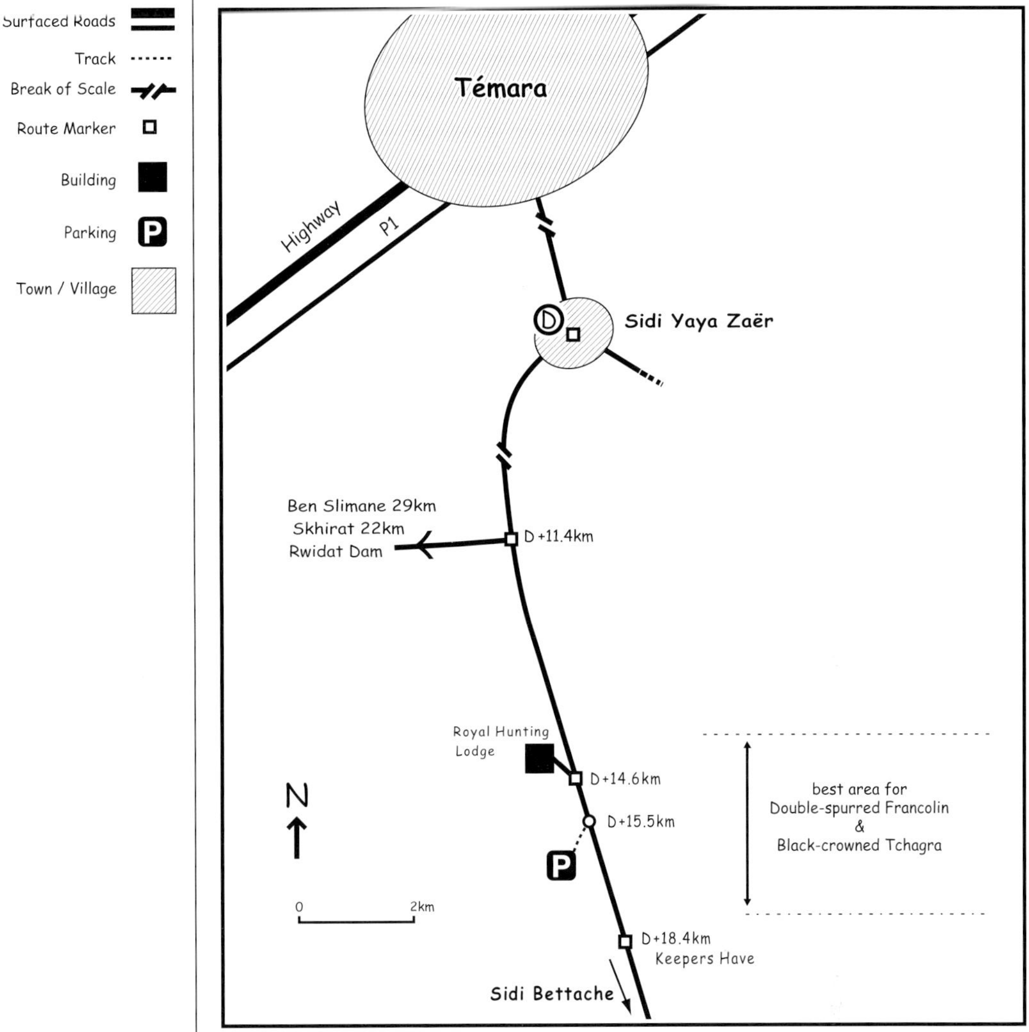

If you stay longer in the region, visit the coast (the estuary of the Yquem Wadi can be good), the famous cork-oak woods of Mamora to the northeast of Rabat, and take the opportunity to go to the Lac de Sidi Bourhaba (see page 37).

Birds

The region is famous for Double-spurred Francolin, which inhabits the undergrowth of the cork-oak woods near Sidi Bettache and Ben Slimane. The forest along the S208 between Sidi Bettache and Sidi Yaya is perhaps the best area to look for this bird which is particularly vocal in the very early morning and most visiting birders stop near the Royal Hunting Reserve, some 15km after Sidi Yaya travelling in the Sidi Bettache direction.

One of the main features of the region is the numerous breeding passerines. In the fields and cultivated areas there are Calandra and Thekla Larks, Stonechats, Zitting Cisticolas, Southern Grey Shrikes, Spanish Sparrows, and Cirl and Corn Buntings. The lovely Moroccan Magpie is fairly common, the Raven much rarer.

Double-spurred Francolin

Rufous Bush Robins, Nightingales and abundant Sardinian Warblers live in the bushes, while Common Bulbuls, and Cetti's and Western Olivaceous Warblers are attracted by the vegetation that fringes the wadis. Great Spotted Woodpeckers, Tawny Owls, European Nightjars, Woodlarks, Spotted Flycatchers, Blue (Moroccan race) and Great Tits, Short-toed Treecreepers, Golden Orioles, Woodchat Shrikes, Black-crowned Tchagras, Hawfinches, and Jays prefer the forests. The much sought-after Levaillant's Woodpecker has spread since the 1980s in the valleys of the Korifla and Cherrat Wadis, and in most forests including the Dar-es-Salam golf course near Rabat.

Cattle Egrets and White Storks are numerous and some Barbary Partridges may be seen. Raptors are fairly common. Black-shouldered and Black Kites, Short-toed Eagles, Marsh and Montagu's Harriers, Sparrowhawks, Long-legged Buzzards, Golden, Booted and Bonelli's Eagles, Lesser Kestrels, Kestrels, Hobbies, and Lanners all breed here. The wadis and their surroundings attract Little Ringed Plovers, Kingfishers, European Bee-eaters, and Red-rumped Swallows.

The birds along the coast include Collared Pratincole, Kentish Plover (resident) and Little Tern which breeds on the beach near Skhirat. Collared Pratincoles sometimes breed in the inland fields. Large numbers of Pallid Swifts gather over the city walls of Rabat at dusk.

A wide range of waders and gulls stop along the shore. A glance at the mouth of the Yquem Wadi is often interesting. Mediterranean and Audouin's Gulls, and Royal and Lesser Crested Terns have been reported, mainly in spring, summer and autumn. Rarities along the shore have included Black-throated Diver, Black-browed Albatross, and Sabine's, Ring-billed and Glaucous Gulls. Andalusian Hemipode was heard singing near Skhirat in April 1988.

Among the species that pass through the region in good numbers are Great Spotted Cuckoo (Jan-Feb), Black Kite (March-April), and European Bee-eater (April-May). A large number of European passerines winter, as do Woodcocks, and Marsh Harriers whose roosts can contain more than 70 birds.

Other wildlife

Several species of amphibians and reptiles live in this region and Horseshoe Snake and Mauritanian Toad are of particular interest. From February to April orchids are in flower: Yellow Bee, Moroccan Bee, Mirror, Sawfly and Moroccan Woodcock Ophrys, Green-winged, Milky and Naked Man Orchids, and Long-lipped and Heart-flowered Tongue Orchids. The Red Deer has been introduced in the Royal Hunting Reserve; other large mammals here include Golden Jackal, Red Fox and Wild Boar.

The region of Doukkala
(3b – Doukkala)

The rich agricultural coastal region of Doukkala, between El Jadida and Safi, holds three interesting sites: the lagoons of Sidi Moussa – Oualidia, the lagoon of Khémis des Zémamra, and Cape Beddouza. The first area has been famous for many years, but the second has only been discovered recently by birders. Cape Beddouza seems to be a promising seawatching site, but little has been done here yet. They are described in the following pages.

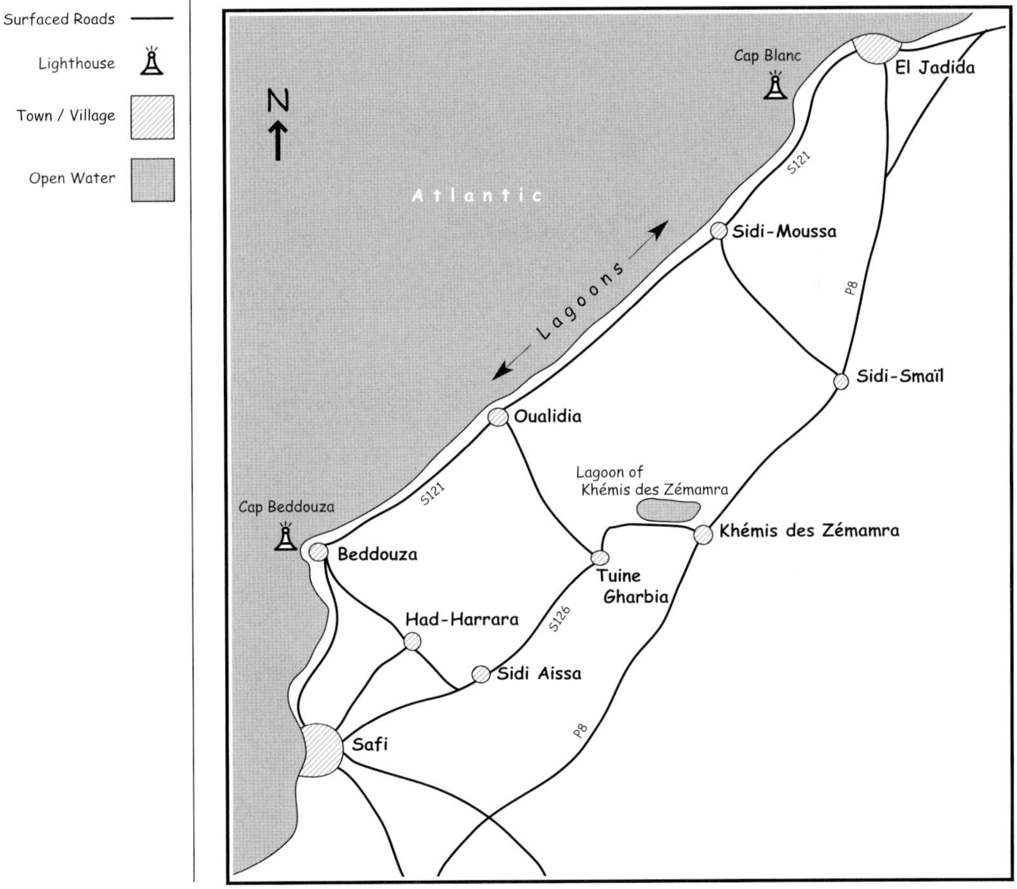

Accommodation

The towns of El Jadida and Safi have a good selection of hotels. Of the two our favourite is El Jadida, a nice, clean and pleasant city, especially when we are travelling south and want to do the lagoons in the morning; good choices include the 'Hotel de Provence' (❷ / ❸ 42 Avenue Fquih M'Hamed Er-Rafy. ☎ 023.34.23.47 and 41.12, Fax 023.35.21.15) and the 'Palais Andalou' (❷ Boulevard Docteur de Lanoë. ☎ 023.34.37.45, Fax 023.35.16.90).

However, we prefer to stay at Oulidia – a popular beach resort which has a few nice hotels/restaurants, all downtown: 'L'Hippocampe' (❶ ☎ 023.36.61.08), 'L'Araignée gourmante' (❷ ☎ 023.36.64.47 Fax 023.36.61.44), the 'Chems' (❷ ☎ 023.36.69.52), and 'L'Initiale' (our favourite in 2002 ❷ ☎ 023.36.62.46); the 'Thalassa' is more basic (❸ ☎ 023.36.60.50) and rents apartments suitable for 4-6 people. Though it is possible to camp in the open near one of the lagoons, it is safer to use the camping site 'Les Dunes d'Or', near the beach at Oualidia.

Oualidia is famous for its oysters, which can be bought direct from the farm. There is also a variety of fish and shellfish available.

The lagoons between Sidi-Moussa and Oualidia

(3b – Doukkala)

This complex system of lagoons and salt pans is an important resting place for numerous migrating birds, and is of special interest for waders. It has been set as a permanent Reserve, protected from hunting since 1984.

Location

The lagoons lie on the Atlantic coast to the southwest of El Jadida, between Sidi Moussa and Oualidia. The lagoons are best explored by car but they can also be easily reached by bus from El Jadida and Safi. Buses travel between these two towns on the S121 which runs alongside the lagoons and will stop near the most interesting birdwatching places. It is also possible to reach the site by hitch-hiking

Strategy

Plan to spend at least half-a-day here, though a full day could be worthwhile. The best times to visit the area are during the peak migration in early spring and autumn. At these times waders are numerous and there are a variety of ducks, raptors and passerines. The lagoons are one of the main wetlands for wintering birds in Morocco and a visit during the winter months is always interesting. Resident passerines such as Crested Lark, Zitting Cisticola, Sardinian Warbler, Goldfinch, and Linnet begin to breed in early March, migrants like Collared Pratincole and Little Tern start later and breeding continues into June. July and August are probably the least interesting months from an ornithological standpoint, but the site is then interesting for swimming and relaxing. While travelling along the S121 the most

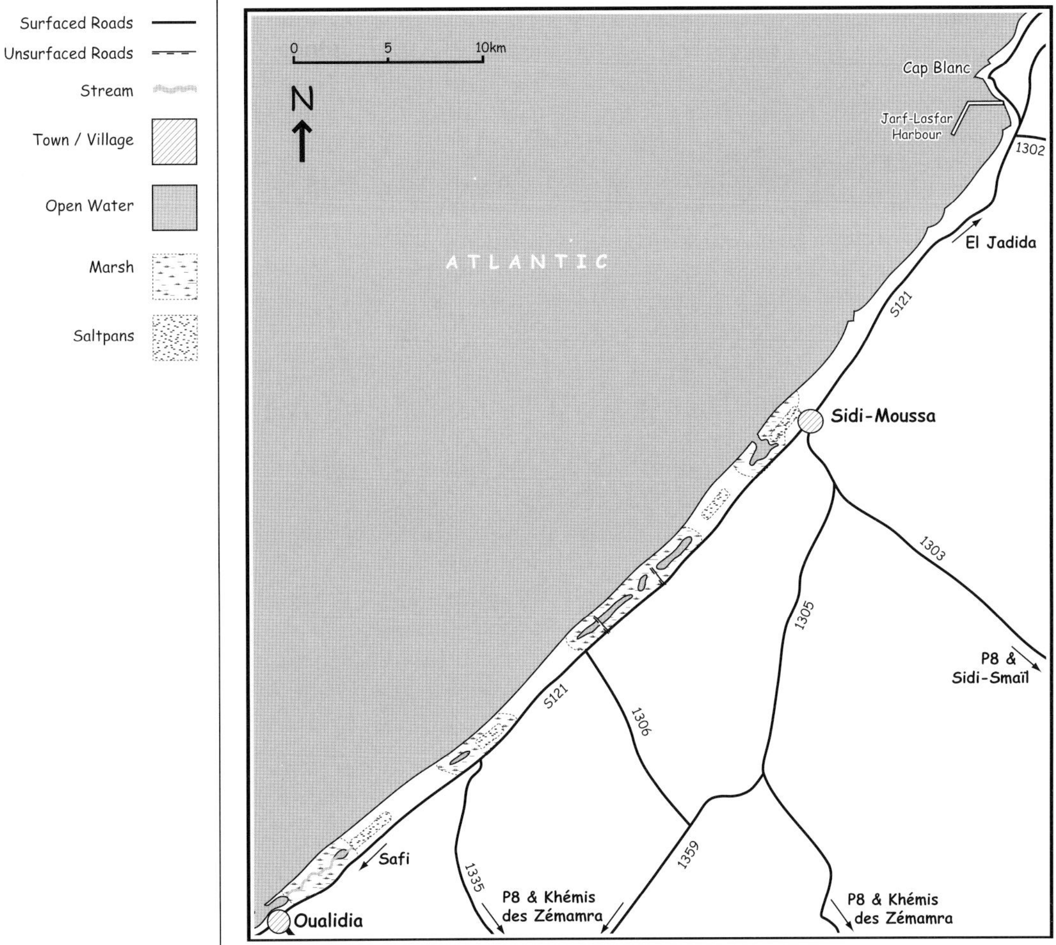

interesting places for birds can be clearly seen from the road. Particularly good places are the lagoons near Sidi-Moussa and Oualidia, and the salt pans between these two localities. Some tracks lead to these areas and allow close approach to the birds. The surrounding land can be explored on foot. The mimosa woods on the dunes between the lagoons and the ocean contain large numbers of breeding finches.

To the north, the cliffs near the harbour of Jorf-Lasfar (Cap Blanc) are worth visiting as this is a good place for species such as Lesser Kestrel, Lanner, Peregrine, and Blue Rock Thrush.

Birds

The number of water birds present at the site varies greatly, from a minimum of 3,000-5,000 in summer (June-July) to a maximum of 15,000-16,000 in October-November. Waders form 60% to 90% of the total, gulls and terns an average of 11%, and ducks 6% to 20%.

Resident species include Eurasian Coot, Moorhen, White Stork, Black-winged Stilt, and Kentish Plover. Collared Pratincole (April-September) and Little Tern (April-October, with some overwintering) are migrant breeders, while Garganey, Red Knot, Curlew Sandpiper, and Bar-tailed Godwit are only passage migrants.

16 other species, which are basically migrants, can also be seen throughout the year: Little Grebe, Little Egret, Grey Heron, Oystercatcher, Avocet, Ringed Plover, Turnstone, Dunlin, Redshank, Greenshank, Curlew, Black-tailed Godwit, Audouin's, Lesser Black-backed and Black-headed Gulls, and Sandwich Tern. Greater Flamingo, Spoonbill, Grey Plover, Sanderling, Little Stint, Spotted Redshank, Marsh Sandpiper, Ruff, Common Snipe, Whimbrel, and Yellow-legged Gull are absent during the summer.

11 species are only winter visitors: Great Cormorant, Glossy Ibis, Shelduck, Wigeon, Mallard, Pintail, Northern Shoveler, Common Teal, Marbled Duck (up to 1,100 in March 1989 and 1,420 in October 1997), Pochard, and Golden Plover. This is one of the major Moroccan wintering areas for Mediterranean Gulls (up to 150 on 25 January 1994) with the Azzemour-El Jadida Bay (238 on 23 January 1992) and the rocky shore of Sidi Bouzid (320 on 4 February 1980), some dozen kilometres to the north.

Occasional visitors include Great Crested and Black-necked Grebes, Shag, Purple Heron, Ruddy Shelduck, Gadwall, Water Rail, Wood, Marsh and Common Sandpipers, Grey Phalarope, Slender-billed Gull, and Gull-billed, Caspian and Black Terns. Finally, many more species have been noted at least once at the site, including Moroccan rarities such as Bulwer's Petrel, Brown Booby, Great Egret, Western Reef Egret Blue-winged Teal, Ring-necked Duck, Goldeneye, Semipalmated, Pectoral and Purple Sandpipers, Long-billed Dowitcher, Lesser Yellowlegs, Laughing and Ring-billed Gulls, and Scarlet Rosefinch. Slender-billed Curlew (30-50 birds in January 1964) has not been recorded since 1993.

Lesser Crested Tern

Seawatching can be rewarding with species such as Cory's, Great, Sooty, Manx and Balearic Shearwaters, and Lesser Crested and White-winged Black Terns or Audouin's Gulls (113 of the latter flew north in 7½ hours of observation on 1 March 1981).

Numerous passerines pass through along the coast; Great Spotted Cuckoo is one of the earliest migrants, passing northwards from

December through March, and southwards from late July through September, while the peak migration periods for most other species are in March-April and September-November. Breeders include Zitting Cisticola and Spotless Starling; the mimosa woods and the adjacent areas attract numerous finches. The secretive Andalusian Hemipode may well survive near the site, in fallow fields and waste lands covered with Dwarf palm; there are 12 records of this species between Azzemour and Oualidia, the most recent one in March 1988.

The lagoon of Khémis des Zémamra

(3b – Doukkala)

This newly discovered site is excellent for ducks and waders and should not be missed if you intend to visit Doukkala. Nevertheless, its future may be endangered as agriculture develops quickly.

Location You cannot miss the lagoon, which is on the north side of the S126 road between Khémis des Zémamra and Tnine Gharbia, 5-6km west of Khémis des Zémamra. It seems to be permanently under water (for example after a severe drought in October 2000 it was still over 1.5km long and 0.5km wide). It is surrounded by cereal, bean, sugar-beet and other fields; a high voltage power line crosses it and cows graze around.

Accommodation No accommodation or food are available at the site; the closest hotels are at Oualidia (see page 44) and it is wise to bring a picnic.

Strategy Plan to spend at least a couple of hours at the site as both the number and variety of birds are generally high. It takes about 45 minutes to drive to the site from Oualidia (c.40km away). The small mounds that have been raised along the S126 give perfect vantage points over the lagoon but a telescope is essential because of the size. The site is very good in spring and autumn and should also be good in winter but the authors do not yet (2002) have any records for this period.

In spring 2001 there were several other lagoons near Tnine Gharbia but none was as interesting as this one. However, the whole region may be well worth exploring. The many little roads that go from the S126 to the coast (only some of them have been drawn on the sketch) cross very peculiar and nice bare limestone landscapes.

Birds Several hundred Little Grebes sometimes occur at the site. Black-crowned Night-Heron, Grey and Purple Herons, Little and Cattle Egrets, Spoonbill, and Greater Flamingo (up to c.1000 on 20 October 2000) have been recorded. The lagoon attracts large quantities of ducks and these have included common European species (Wigeon, Gadwall, Pintail, Common Teal, Garganey, Mallard, Northern Shoveler) as well as some more interesting species such as Ruddy Shelduck (up to 22 on 1 September 1999) and Marbled Duck (up to 16 on 8 April 1999).

Waders are often very numerous and a large variety has been recorded here: Black-winged Stilt, Avocet, Collared Pratincole, Little Ringed, Ringed and Kentish Plovers, Northern Lapwing, Red Knot, Little Stint, Curlew, Green, Wood, Common and Marsh Sandpipers, Dunlin, Sanderling, Ruff, Common Snipe, Black-tailed Godwit, Redshank and Spotted Redshank, Grey Plover, Greenshank, and Turnstone. It is likely that rare species will be found here in the future!

In migration periods, Gull-billed and Whiskered Terns come to feed at the lagoon. In the nearby fields Quail, Calandra Lark, Yellow Wagtail, and Zitting Cisticola may be recorded.

Cape Beddouza
(3b – Doukkala)

This is quite probably a very good Moroccan seawatching spot but no extensive birding has been done yet. We welcome any new observations!

Location

The Cape is mid-way from Oualidia and Safi on the coastal road S121.

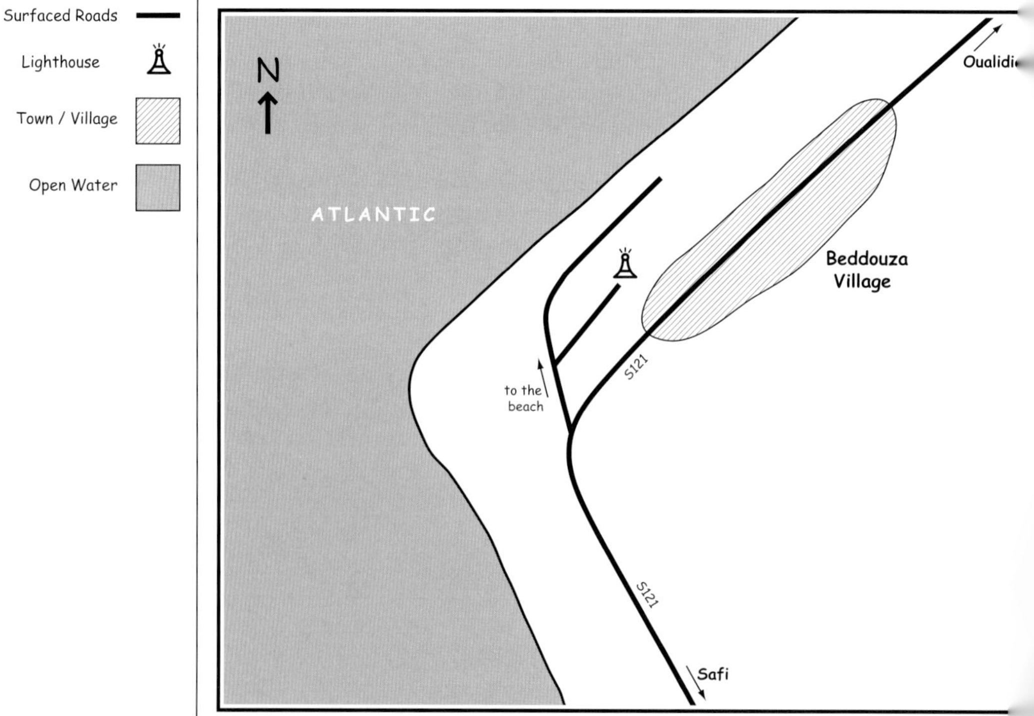

Accommodation

The 'Auberge Cap Beddouza' was the only accommodation availab at the site in 2001 but it was in such a poor condition then that it was wiser to sleep at Oualidia.

Strategy Coastal mist often prevents any observation before noon and we recommend visiting here in the afternoon. A small tarmac road starts near the lighthouse and allows closer views of the ocean. Also take a look at the nearby sea cliffs as they are good for species such as Lanner, Lesser Kestrel, Blue Rock Thrush, and Raven.

Birds Among the shearwaters, Cory's is an abundant passage migrant mainly in October-November and February-May, Balearic is fairly common in September-October and February-March and Great, Manx, and Sooty are all much rarer. European Storm-petrels may be seen mainly in August-November; Leach's Storm-petrel is rarer. There is one record of Sooty Shearwater (9 on 17 October 2001), Bulwer's Petrel (2 July 1972) and Madeiran Storm-petrel (10 December 1989).

Passages of Northern Gannets can be impressive (up to 1300 birds per hour on 31 October 1962); Cape Gannet has been claimed but there is no proven record as yet. Arctic Skua is a common autumn passage migrant and uncommon winter visitor, Great Skua is rarer and Pomarine is a scarce migrant and an occasional winter visitor. Common Scoter can be fairly common at sea from mid September through April.

Gull and tern migrations can also be rewarding. The commonest is probably Sandwich Tern but many other species may be sighted at the Cape including Mediterranean and Audouin's Gulls and Caspian, Lesser Crested, and Royal Terns. A Glaucous Gull was seen on 17 November 1990.

A Bald Ibis colony was established here in the early 1950s; but is only a memory now. A juvenile Allen's Gallinule was recorded on 2 January 1989.

Essaouira

(3d – Chiadma)

The delightful town of Essaouira with its narrow streets, cottage industries (especially cabinet making and inlaid juniper) and fishing harbour, is the best site in Morocco to see the graceful Eleonora's Falcon which is present between May and October.

Location Essaouira is located on the Atlantic coast 173km north of Agadir and 176km west of Marrakech. The city can be easily reached by bus.

Accommodation Within the last few years, Essaouira has become one of the most popular towns in Morocco, and tourism has developed dramatically. A lot of new hotels have been opened to cater for the tourist trade. Some are located near the ocean, e.g. the 'Tafoukt' (❶ 58 Boulevard Mohamed V. ☎ 044.78.45.04 and 044.47.25.05, Fax 044.78.45.05), while others are in the old town: the 'Souiri' (❸ / ❷ 37 rue Laatarine. ☎ and Fax 044.47.53.39), the 'Sahara' (❷ Avenue Okba Ibnou Nafiaa. ☎ 044.47.52.92), or the 'Maison du Sud' (❶ 29 avenue Sidi Mohamed

Ben Abdallah. ☎ 044.47.41.41, Fax 044.47.68.83). A selection of hotels is available on http://www.essaouiranet.com/.

Outside of the town, we recommend the 'Auberge Tangaro' (❶ Quartier Diabat. ☎ 044.78.47.84), c.10km after the exit of the town on the P8A, toward Agadir (in the middle of juniper forest with great views of the bay and the islands).

Some restaurants serve excellent fish and sea-food, and the open-air ones, near the harbour entrance are also worth trying, even though the are often not good value for money.

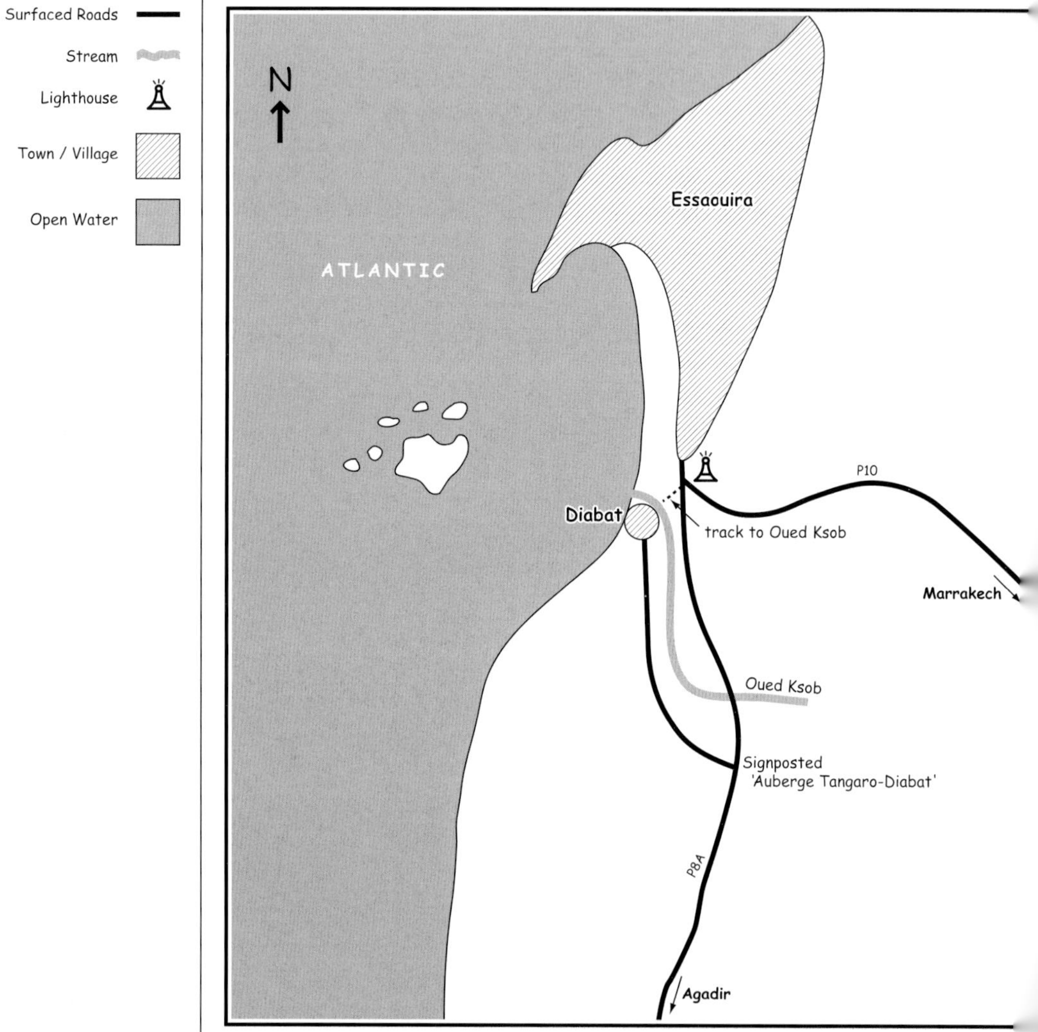

Strategy

The Eleonora's Falcons breed on the islands offshore and the archipelago has been set up as a reserve with prohibited access. The best vantage point for viewing the islands is the end of the jetty on th west side of the harbour, and a telescope is essential. The small islets immediately to the west of the harbour, the harbour itself, and the ba are worth watching for seabirds.

A good place to look for the falcons when they are hunting is the mouth of the Oued Ksob a few miles to the south which is also a roosting place for waterbirds. The northern bank of the Oued is reached by a track that runs through the tamarisks and junipers; it starts some 100 metres after the lighthouse on the P8A, just after the exit of the town, and it is signposted 'Diabat Café / Restaurant 500m' (the Café is actually on the other side of the Oued, in Diabat, which became famous as a hippy commune with residents such as Jimmy Hendrix, but has now reverted to an ordinary village). The southern side can be reached via the road for the 'Auberge Tangaro', starting on the P8A 3.8km south from the lighthouse. Allow for a few hours for the falcons and the mouth of the Oued Ksob.

Birds

One of the best times to visit Essaouira is May and June. The Falcons, which arrive in late April, have not yet begun to breed and so do not spend so much time on the islands. The population in the mid 1990s was about 230 pairs then dramatically increased in the early 2000s, after severe protection measures; 675 breeding pairs have been recorded in 2000 and 2001, 10% of the world population, making this colony one of the largest in the world. The other breeding birds of the islands are c.2,500-3,000 pairs of Yellow-legged Gulls, some tens of pairs of Moroccan Great Cormorants, one pair of Peregrines, Kestrels and Common Ravens, numerous Rock Doves, and several pairs of Pallid Swifts and Sardinian Warblers.

The mouth of the Oued Ksob is a morning and evening resting place for Yellow-legged Gulls, and a wide range of other birds stop there, including Audouin's Gull, and Sandwich, Lesser Crested, Little, Whiskered, Black and White-winged Black Terns. Two nearctic Gulls have been seen here: Ring-billed and Glaucous-winged, and Western Reef Egret has been reported once.

The Oued also attracts several waders, and Grey Herons, Little Egrets, Spoonbills, Avocets, Little Ringed Plovers and Greenshanks have been noted there. Plain Martin and possibly Kingfisher breed in the banks, while Moroccan White Wagtail (*subpersonata*) is a common sight. Cetti's Warblers can be heard in the bushes all along the Oued.

In the scrub around the mouth of the wadi and in the Juniper woodland along the P8A and the P10 there are Barbary Partridge, Quail, Stone-curlew, Turtle Dove, Red-necked Nightjar, Common Bulbul, Rufous Bush Robin, Moussier's Redstart, Black-eared Wheatear, Olivaceous and Sardinian Warblers, Spotted Flycatcher, Moroccan Blue Tit, Great Tit, Southern Grey and Woodchat Shrikes, Moroccan Magpie, Chaffinch, Goldfinch and Serin. Great Spotted Cuckoos are regular in spring migration, between December and April.

Pallid Swifts, House Buntings, Spotless Starlings, Collared Doves, and occasionally Eleonora's Falcons can be seen while walking round the city.

Other wildlife

In the surrounding countryside Algerian Hedgehog, Brown Hare, and Wild Boar are not uncommon. Genets are sometimes seen and Spur-thighed Tortoise is common.

N'Fiss Reservoir (Lalla Takerkoust Barrage)

(3i – Haouz)

N'Fiss Reservoir is a large reservoir built on the Oued N'Fiss close to Marrakech, one of the most popular tourist destinations in Morocco. Also called Lalla Takerkoust Barrage, it is often good for birds and well worth a visit when spending time in Marrakech.

Location

The reservoir lies 30km south-southwest of Marrakech and is easily reached following the S507 road. The drive takes half-an-hour or so and the road is well sign-posted starting from Marrakech.

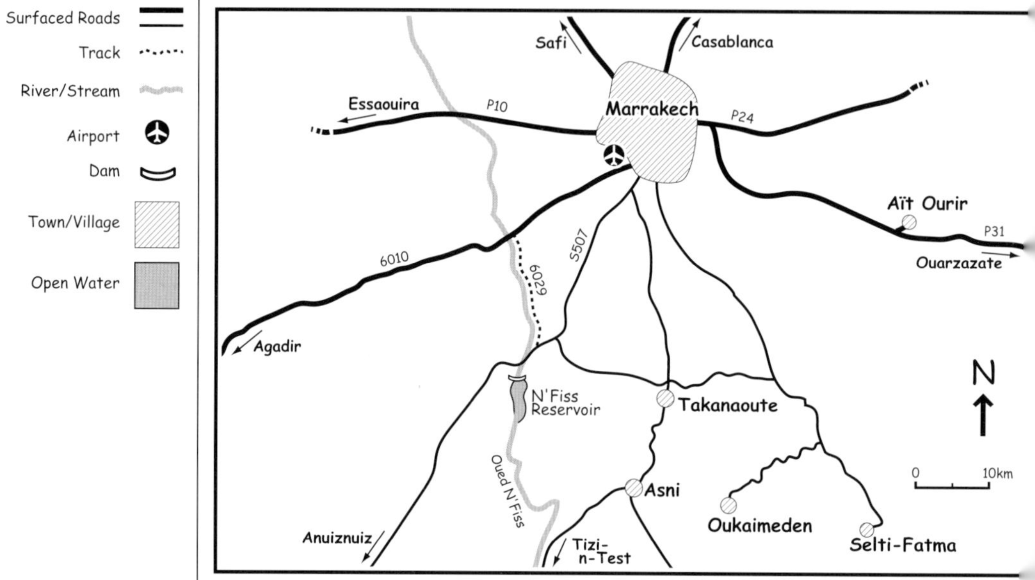

Accommodation

There is no accommodation available near the site itself, but it is close to Marrakech where a whole range of hotels is available. It is also close to the northern foothills of the High Atlas, near Asni for example (see 'The Tizi-n-Test Road' and 'Imlil' page 62 and 65). The restaurant "Relais du Lac" overlooks the western bank of the reservoir and seemed promising in 2001.

Strategy

The dam (357m long, 62m high and 47m wide at the base) was built on the Oued N'Fiss to provide water for the Haouz Plain and Marrakech city. When full of water it forms a 7km-long reservoir with a capacity of 78 million cubic-metres but the actual capacity depends much on seasons and years. Nevertheless it is often attractive for birds especially during spring and autumn.

The best access and viewpoints are from the western side of the reservoir. Turn right before the village of Lalla Takerkoust, cross the Oued N'Fiss on the bridge that has been built just below the dam and then turn left to the 'Le Relais du Lac' restaurant. Several vantage points are available along this 5km-long tarmac road.

The access to the eastern bank is on a good track that passes through the village of Lalla Takerkoust. It is not so easy but can be quieter.

Because of the size of the site, a telescope is essential to get good views of the birds.

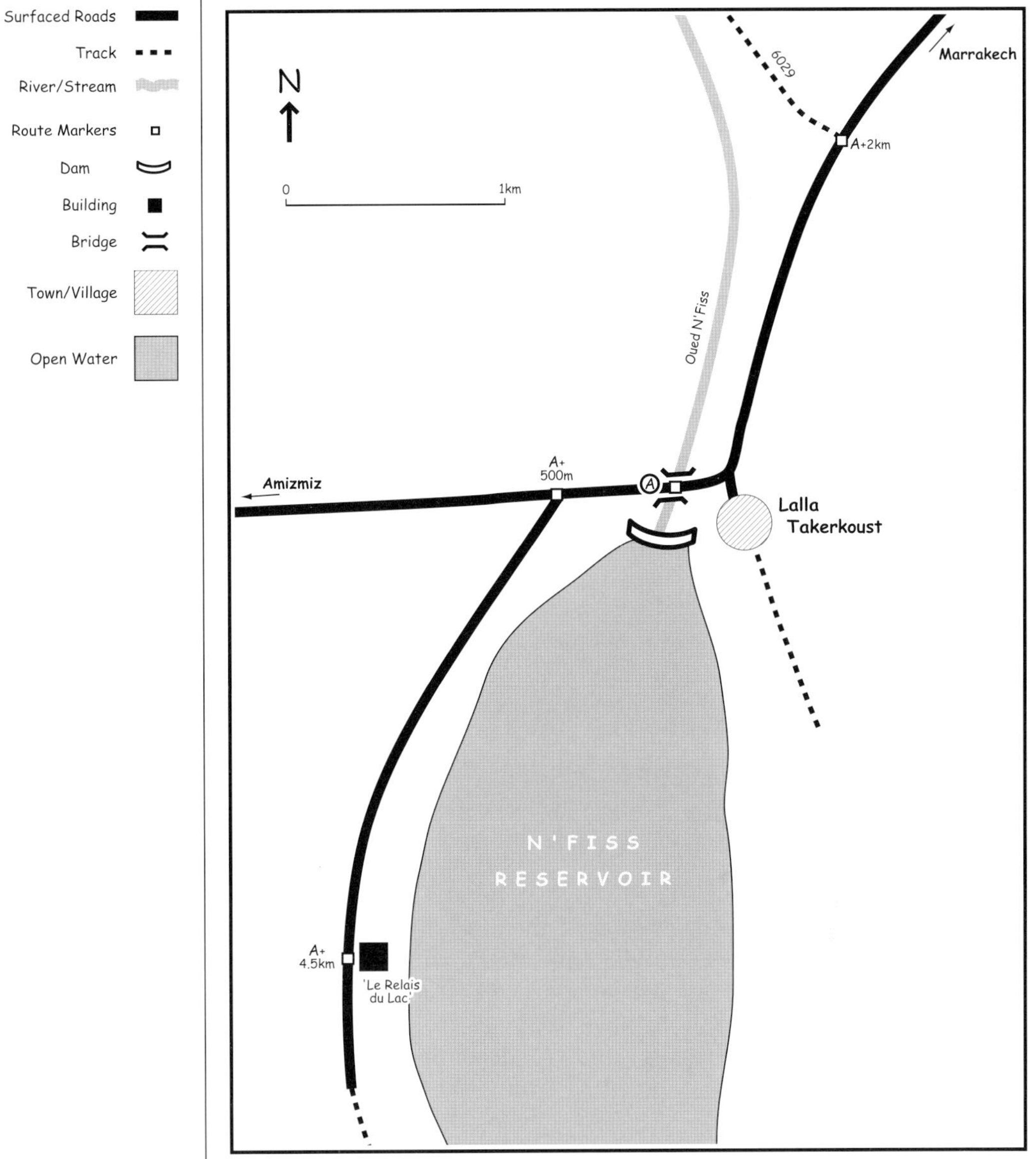

An alternative circuit back to Marrakech is to follow the 6029 road that skirts the right bank of the Oued N'Fiss. It starts some 2km downstream from the dam and is signposted 'Agadir Tachraft' (in 2001 the first 11km was a good track, then the remaining 5km was tarmac). It joins the 6010 road that runs from Marrakech to Guémassa/Agadir. Possible species here include Peregrine, Lanner and Black Wheatear. Bald Ibis once nested in the cliffs near the Oued. The largest colony was at Agadir Chems some 10km downstream from the dam (up to 15 nests were recorded in 1975; last record in 1976) and another one (a few pairs) nested closer to the barrage from 1979 to 1986; there have been no records since the end of the 1980s.

Birds

The authors have recently described the avifauna of the Marrakech region, including N'Fiss Reservoir in a paper in the French journal *Alauda,* so those requiring detailed information on the birds of this region should refer to that (see bibliography). Copies are available from the authors.

In spring, stop on the roadside along the Oued N'Fiss, a few hundred metres before reaching Lalla Takerkoust. Here good views can be had over the Oued and the surrounding thick vegetation which shelters Moorhen, Moroccan White Wagtail, Common Bulbul, Nightingale, and Cetti's and Western Olivaceous Warblers. A colony of Cattle and Little Egrets and Black-crowned Night-Herons is established on the tall trees.

A lot of waterbirds occur on the reservoir in migration periods and in winter, and the most commonly recorded include Black-necked Grebe, Grey and Purple Herons, White and Black Storks, Spoonbill, Shelduck, Wigeon, Common Teal, Mallard, Pintail, Northern Shoveler, Pochard, Avocet, Ringed and Little Ringed Plovers, Dunlin, Little Stint, Ruff, Common Snipe, Black-tailed Godwit, Redshank, Spotted Redshank, Greenshank, and Green and Wood Sandpipers. Accidental visitors have included Greater Flamingo, Greylag Goose, Gadwall, Garganey, Marbled and Ferruginous Ducks, Black-winged Stilt, Golden and Grey Plovers, Northern Lapwing, Red Knot, Sanderling, Temminck's Stint, Curlew Sandpiper, Curlew, and Grey Phalarope. A Blue-winged Teal, a real Moroccan and Western Palearctic rarity, was seen here in October 1984.

Ruddy Shelducks are often present in good numbers and up to 150 were recorded in October 1981; some pairs may breed. The Kentish Plover is also an occasional breeder.

Black-headed Gull, and Whiskered and Black Terns are commonly recorded while Lesser Black-backed Gull and Gull-billed Tern are accidental.

Long-legged Buzzard, Booted Eagle, Osprey, Kestrel, Lanner and Peregrine are among the most often recorded raptors. Little Owl is fairly common. Long-eared Owl has bred in the eucalyptus woods. European Bee-eater is a common breeder while Roller and Hoopoe are rarer. Rufous Bush Robin, Black and Black-eared Wheatears, and Southern Grey and Woodchat Shrikes are among the most interesting passerines breeding around the reservoir.

The region of the Lakes
(5a – Western Middle Atlas)

This area to the northeast of the Middle Atlas contains a cross-section of interesting habitats such as the cedar woods, the stony plateaux ('causses') and the lakes (called 'dayets'). Several endemic subspecies and species and some Moroccan rarities can be found here.

Location

The region is located some 50km south of Fès. The city (and skiing resort) of Ifrane, a good starting point for trips, can be easily reached by bus, but most of the region is best explored by car.

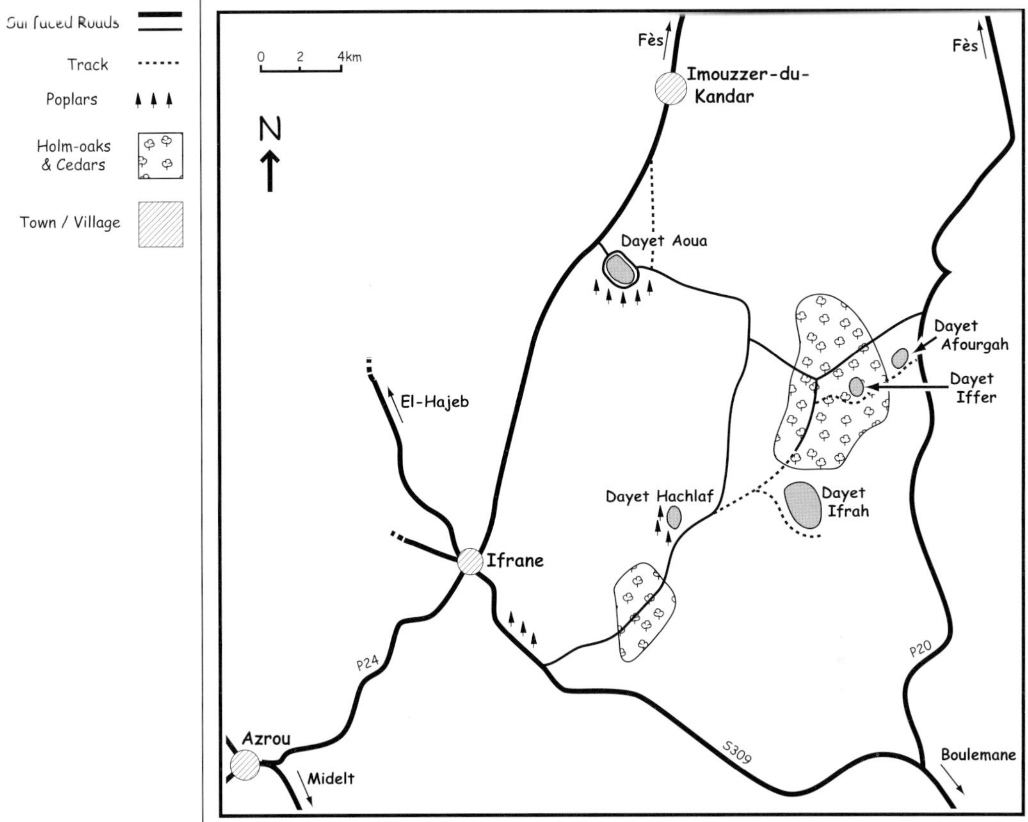

Accommodation

Ifrane is probably the best base and has some upper category hotels: the 'Michliffen' (❶ ☎ 055.56.66.07, Fax 055.56.66.23), the 'Perce Neige' (❶ Rue des Asphodèles. ☎ 055.56.63.50, Fax 055.56.71.16), and the 'Tilleuls' (❷ ☎ 055.56.66.58). An alternative base is the charming little town of Azrou although it is situated 17km to the southwest of Ifrane and thus further away from the birding sites. Hotels here include the 'Panorama' (❷ Rue El Hansali. ☎ 055.56.20.10, Fax 055.56.18.04) and the cheaper 'Des Cèdres' (❸ Place Mohammed V. ☎ 055.56.23.26), and 'Azrou' (❸ Route de Khénifra. ☎ 055.56.21.16). The city of Imouzzer du Kandar, 25km to the north of Ifrane, can also be a good starting point, with the 'Charazed' (❷ 2 place du marché. ☎ 055.66.36.70, Fax 055.66.34.45) and the 'Royal' (❷ Boulevard Mohammed V. ☎ 055.66.30.80). Other good alternatives are near Dayet Aoua: the 'Gîte du Lac Dayet Aoua' (❷ ☎ 055.60.48.80, Fax 055.60.48.52, e-mail aouagite@yahoo.com) and the 'Chalet du Lac' (❷ ☎ 055.66.31.97).

Strategy

The northeast Middle Atlas is generally under snow from late December to February making the tracks and sometimes the roads impassable but this is a good period for ducks and resident passerines (you can also go skiing at Ifrane). The most pleasant period is from April to July; August and September are quieter months but a visit then provides an escape from the heat of the surrounding plains.

The region can be explored by car in half a day though it can be worth staying longer in the region, especially when it is hot in the surrounding plains. Most of the tracks listed in the 1st edition of this guide have now been asphalted, some others are drivable but are poorly marked; however there is little risk of getting lost. An interesting circuit consists of visiting the three main lakes Dayets Hachlaf, Ifrah, and Aoua (see map), stopping to explore the various habitats en route. Dayet Iffer offers great scenery, but holds very few birds; the vegetation around Dayet Afourgah has been cut, and the lake is now much less interesting. If you have more time to spare then walking and camping is a good way to see the wildlife.

To the northwest of Ifrane the road to El-Hajeb (S309) passes through the forest of Jabaa (beautiful oak woodland, good for birds).

To the south of Ifrane the lake of Aguelmane Azigza, above Khénifra is worth visiting for the splendid scenery. If travelling on to the upper valley of the Oued Moulouya (page 77) stop at Foum Kheneg gorge on the P21 south of Timahdite, and Aguelmane Sidi Ali near the Zad pass, for the abundant Ruddy Shelducks.

Birds

The wealth of the lakes depends much on the surrounding vegetation and whether it has been cut or not (in the 1980s, Dayet Afourgah held a colony of herons with breeding Black-crowned Night-Heron, Cattle and Little Egrets, and perhaps Purple Heron, but the reeds have been cut since). The Dayets Hachlaf, Ifrah and Aoua are now probably the most interesting of the main lakes. In May and June you can see or hear Great Crested Grebe, Little Bittern, Marsh Harrier, Eurasian and Red-knobbed Coots, Black-winged Stilt, and Savi's and Great Reed Warblers. During the migration period and in winter, the lakes sometimes hold large quantities of ducks; 1,200 Marbled Teals were at Dayet Aoua in April 1999. The lakes are one of the best Moroccan places for Red-knobbed Coots, which can breed as early as December (24 December 1992) and as late as September (young chicks early October 1997); Dayet Aoua is enormously important for this species.

Numerous passerines breed in the cedar and holm-oak woods and include endemic subspecies of Mistle Thrush, Firecrest, Coal Tit, Great Tit, Nuthatch, Short-toed Treecreeper, Jay, Chaffinch and Hawfinch. This is one of the best places for other interesting birds such as the North African race of Great Spotted Woodpecker, Stock Dove, Moussier's Redstart, Western Bonelli's Warbler, Golden Oriole, and Cirl Bunting.

The recently-split Atlas Flycatcher (*Ficedula speculigera*) arrives from late April and leaves by end August; it favours mature forests of Atlas cedar and oaks. The local race of Blue Tit (*ultramarinus*) is an abundant resident, and is also a good candidate for treatment as a separate species. The poplars that fringe some rivers and lakes (especially by Dayet Aoua and Hachlaf, or along the S309 a few kilometres to the southeast of Ifrane) are excellent for Levaillant's Green Woodpecker.

Raptors include Black and Red Kites (the latter on the verge of extinction), Short-toed Eagle, Sparrowhawk, Goshawk, Long-legged Buzzard, Booted Eagle and Hobby. Egyptian Vulture has become very rare after wide campaigns of poisoning. The Red-footed Falcon, a moroccan rarity, has been recorded at Dayet Aoua (22 April 2000).

Birds which can be found on or near the stony plateaux include Lesser Kestrel, Little Owl, Roller, Thekla Lark, Woodlark, Skylark, Horned Lark (endemic subspecies), Tawny Pipit, Black-eared and Black Wheatears, Melodious and Subalpine Warblers, Spotless Starling, and Rock Sparrow. One of the commonest birds is the northwest African race of Northern Wheatear (Seebohm's Wheatear). Ravens can gather in large groups, e.g. 200+ on 24 September 1999.

Sadly the tiny relict population of Demoiselle Cranes has disappeared (last breeding attempt in 1984). It is worth going out at night to listen for Scops, Eagle, and Tawny Owls, and European Nightjar.

As early as September the first passage migrants arrive and from November all the wintering species can be observed (in particular European ducks).

Other wildlife

European Pond Terrapins live in the lakes, Iberian Wall Lizards, which are very variable in colour, are abundant on the plateaux, and mammals such as Jackal, Fox and Weasel are common. Groups of Barbary Apes are regularly seen in the cedar woods near Ifrane-Azrou. In May and June a lot of orchids flower in the region, including Pink Butterfly, Bug, Robust Marsh, and Algerian Butterfly Orchids.

Tamri and Cape Rhir

(6a – Haha)

The region of Haha is the westernmost part of the High Atlas and contains splendid mountain and ocean scenery. Tamri holds one of the few remaining colonies of Bald Ibis and Cape Rhir is one of the most famous seawatching spots in Morocco.

Location

The area is located along the main P8 road some 50-60km to the north of Agadir (c.45 minutes drive) and 120-130km to the south of Essaouira (three hours straight drive, but generally much longer because of the truck traffic and the various stops en route for birds).

Accommodation

No accommodation is available at Tamri, though there are a lot of small open-air restaurants (including the 'Café Ibis chauve' – Bald Ibis) offering traditional tajines. The wonderful local small bananas are also worth trying. The closest hotels are at Taghazoute: 'Résidence Amouage' (❷ ☎ and Fax 048.20.00.06) and 'Auberge du Littoral' (❷ ☎ 048.31.47.26, Fax 048.31.43.57) but it may be better to stay in Agadir which is a strategic location for visiting various sites of the region (see 'The National Park of Souss-Massa', page 81). The hotels of the Paradise Valley (see page 60) may also be a good choice.

There are a couple of basic camp sites but there is no difficulty with camping in the wild, although it is wise to stay near other campers for safety reasons.

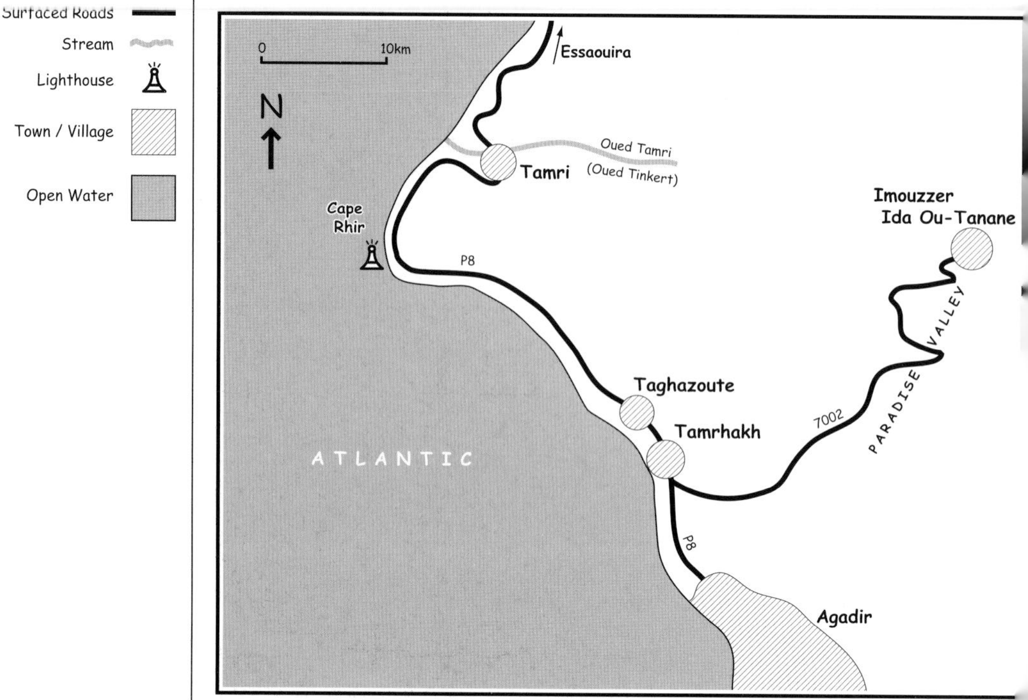

Strategy

If based in Agadir, this site can be visited in half-a-day, though a full day is worthwhile. En route the road passes through or near argan and thuya woodlands, several beaches (good for gulls and terns), coastal cliffs (Plain Swifts have been reported 3km to the north of Taghazoute), Cape Rhir (seawatching), small hills (feeding area for Bald Ibis), the mouth of the Oued Tinkert (also called Oued Tamri; good for Ibises, gulls and terns), the village of Tamri and then other coastal cliffs (also good for the Ibis).

Young 'guides' will probably offer to show you the Ibis and this may be helpful but beware that some of these youngsters are so keen that they often flush the birds. Always explain the correct approach to watching this highly endangered species, and try to keep the 'guides' under control.

Birds

Seawatching is best from the Cape. Cory's Shearwater can be abundant and records have included 21,000+ *borealis* moving north during 6-23 August 1983 and 5,000 *diomedea* in 1.5 hours on 1 May 1998; this last race passes mainly in October-November and then in February-May. Great and Sooty Shearwaters are much rarer and have only been recorded in autumn (August-November). There is only one spring and one winter records of Sooty. Manx is uncommon along the Moroccan Atlantic coast and has only been seen once here (18 February 1999) but Balearic is fairly common in September-October and February–March.

European Storm-petrel occurs mainly in August-November; Leach's Storm-petrel is rarer. Other rare or accidental species have included Fea's/Zino's Petrel (21 August 1983), Bulwer's Petrel (25 May 1961 and 6 August 1983), Little Shearwater (2 possible records on 7 and 16

August 1983) and Madeiran Storm-petrel (27-29 November 1985).

Movements of Northern Gannets can be impressive from August through to mid-November (peaking in October) and in March-May. Common Scoter winters at sea off the Cape. Pomarine and Arctic Skuas pass by in August-October and March-May; Great Skua is rarer.

Passages of gulls and terns can also be heavy with one flock of more than 5,000 Black Terns seen resting on the sea near Taghazoute on 19 April 1999. Among the commonly reported species are Black-headed, Lesser Black-backed, and Yellow-legged Gulls, and Sandwich Tern but many other species may be sighted from the Cape or resting on the beaches and these include Mediterranean and Audouin's Gulls – e.g. c.850 Audouin's on 16 September 1999 and c.400 on 2 October 1999 between Tamrhakh and Tamri, and Gull-billed, Caspian, Lesser Crested, and Royal Terns.

Accidental species for Morocco have included Red-throated Diver (30 March 1993), Black-browed Albatross (17 March 1983 and 15 October 2001), Glaucous Gull (5 April 1977), and Guillemot (23 December 1983, 6 December 1985 and 2 March 2002).

The mouth of the Oued Tinkert is a good resting place for Bald Ibis, especially around the middle of the day. In recent years, up to 65 birds have been seen in one day – 2 October 1999. The fields north of the Cape Rhir lighthouse and the steppes on the south and north sides of the estuary are also good. The breeding population ranged from 19 to 65 pairs between 1994 and 2001, and 160 young were raised (1994-2000).

Bald Ibis

The mouth is also good for gulls and terns, but not so good for waders and only a few are usually present. However, a visit here is always worthwhile as accidental species may be found (Western Reef

Egret, Sabine's Gull, Roseate Tern, Short-eared Owl and Richard's Pipit have all been reported) among more regular species such as Little Egret, Spoonbill, Osprey, etc. Zitting Cisticola is a common breeder in the nearby halophytic vegetation.

The sea-cliffs are home for two puzzling species. The '*atlantis*' Peregrine seems to be an intermediate form between the Mediterranean race *brookei* and the African race *minor*, and this region is one of the few contact zones between this form and the closely related Barbary Falcon – thus careful identification is essential. A small black swift species has been reported several times since 1988 around the cliffs north of Taghazoute (up to 50 pairs in April 1990) when breeding was likely to have been taking place; they have been claimed as Plain Swifts and the Moroccan Rare Birds Committee would very much appreciate full details of any further sightings.

These cliffs are also one of the last sites for the scarce local race of Shag (*riggenbachi*); Lanner and Common Ravens also breed.

The scrubs, argan and thuya woods are the habitat of Barbary Partridge, Little Owl, Thekla Lark, Rufous Bush Robin, Moussier's Redstart, Black-eared Wheatear, Sardinian, Spectacled and Orphean Warblers, Black-crowned Tchagra, Southern Grey and Woodchat Shrikes, and Serin. It is also worth studying the local races of some common European birds: Blackbird (*mauritanicus*), Great Tit (*excelsus*), Magpie (*mauritanica*), Chaffinch (*africana*), Goldfinch (*parva*), and Greenfinch (*voousi*). In late winter and early spring migrant Great Spotted Cuckoos may be seen.

Other wildlife

The cape is good for sea mammals, e.g. Common and Bottlenose Dolphins, and Killer and Fin Whales have all been recorded here on 30 April 2000.

Paradise Valley

(6a – Haha)

This is a very pleasant area in the westernmost Atlantic part of the High Atlas, especially when the weather becomes too muggy near the ocean. Paradise Valley is also a good alternative for accommodation if you don't like the popular, crowded town of Agadir and has some famous restaurants. In addition, the area holds some interesting birds.

Location

Paradise Valley (road 7002) stretches for some 50 kilometres between Tamrhakh (12km to the north of Agadir on the main P8 road) and the village of Imouzzer Ida Ou-Tanane which currently (2002) marks the end of the asphalted road. The track 7003 that starts before Imouzzer and joins the Agadir-Marrakech main road P40 will possibly be asphalted in coming years.

Accommodation

Two good hotels were available in the valley in 2002 and both of them have a swimming-pool. The 'Tifrit' has been built under huge olive trees 32km from Tamrhakh; it was good value for money

(❷ ☎ 048.82.60.44). The 'Cascades' is located at Imouzzer itself, in splendid setting and with a tennis court, but is more expensive (❶ ☎ 048.82.60.16 and 60.23, Fax 048.82.60.24). The restaurant 'A la Bonne Franquette' (☎ 048.82.31.91, GSM 061.28.48.73) planned to expand and offer accommodation with a few high standard rooms.

Several small restaurants in the valley serve traditional salads and tajines but three are especially recommended; the 'Tifrit' for succulent Moroccan specialities, the 'A la Bonne Franquette' for its wonderful French food (both of them very good value for money), and the superb, high-standard restaurant of the hotel 'Cascades'.

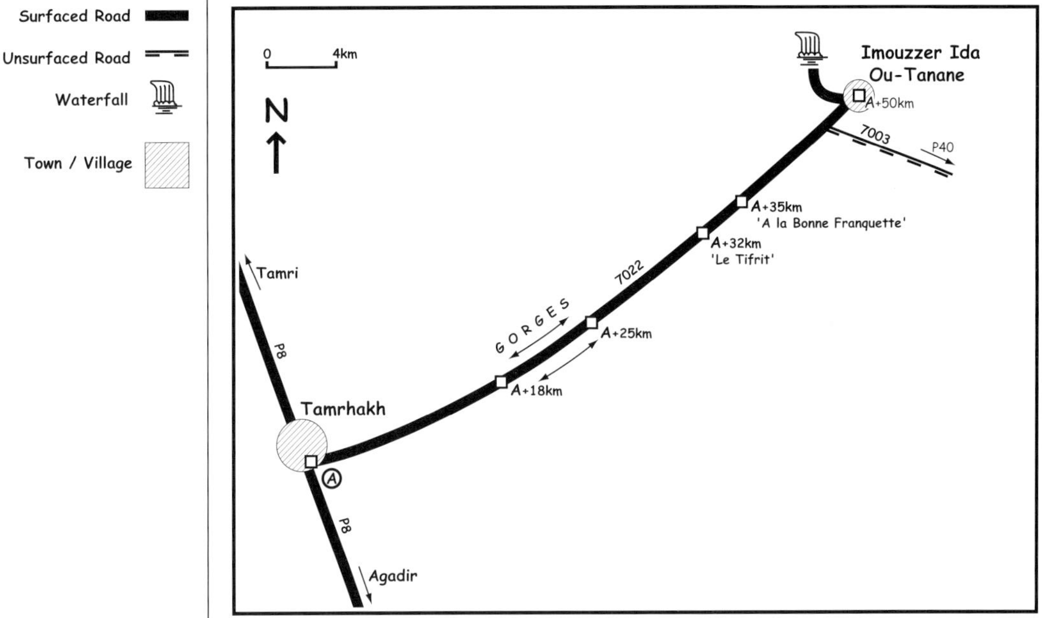

Strategy

The valley can be visited in one full day starting from Agadir but consider spending longer and staying overnight, especially during the hotter months. Almond trees are in full bloom in February and form an unforgettable sight. It is wise to avoid Sundays when many inhabitants of Agadir drive up the valley. Thursdays are interesting for the souk (open market) at Imouzzer.

Several habitats can be explored en route; the argan and thuya woods, the oases with palm and banana trees in the impressive gorges of the Assif (= Oued) Tamrhakh (between 18 and 25km from Tamrhakh village), the olive and almond orchards, the scrubs at high altitude (1,300m) and then the gardens, palm trees and Aleppo pines near Imouzzer. The waterfalls downstream from Imouzzer are also worth a visit although they are often very crowded.

Dealers along the road will offer to sell you many fossils but be careful as most of them are fake. Those wishing to buy a souvenir, would do better choosing the local argan oil or local honey.

Birds

Spotless Starlings and Collared Doves are common near the ocean at Tamrhakh; neither are found higher in the valley.

The argan and thuya woods interspersed with dwarf palms are the habitat of Barbary Partridge, Little Owl, Thekla Lark, Rufous Bush Robin, Moussier's Redstart, Black-eared Wheatear, Sardinian and Orphean Warblers, Black-crowned Tchagra, Southern Grey and Woodchat Shrikes, and Serin. It is also worth studying the local races of some common European birds: Great Tit (*excelsus*), Magpie (*mauritanica*), Chaffinch (*africana*), Goldfinch (*parva*), and Greenfinch (*voousi*).

Other species in the gorges of the Assif Tamrhakh include Turtle Dove, Little Swift, House Martin (several colonies in the cliffs), Common Bulbul, Nightingale, Western Olivaceous Warbler, Spotted Flycatcher, Blue Tit, House Sparrow, and House Bunting. Kingfisher possibly breeds. Grey Wagtail, Black Redstart and Song Thrush are winter visitors.

The cliffs are excellent for Blue Rock Thrush and Black Wheatear. On the 15 April 2001 we also had two Bonelli's Eagles (one overflying the gorges and another just after the Hotel Tifrit); a Red-rumped Swallow was drinking at the swimming pool of the Hotel Tifrit, and several others were overflying the Hotel des Cascades.

The high altitude scrubs are the habitat for Common Cuckoo, Tristram's Warbler, Linnet and Rock Bunting. The Aleppo pines near Imouzzer are home for the local race of Crossbill (*poliogyna*). Accidental species have included Dunnock, and Pallas's and Yellow-browed Warblers.

Other wildlife

The North African Green Frog is common wherever there is water; you will also see several Bibron's Agamas sunbathing along the road.

The Tizi-n-Test road
(6b – Western High Atlas)

This is one of two only roads that cross the central part of the High Atlas range, the other one being the Tizi-n-Tichka road (see Page 69). It runs from the Souss region northeast to Marrakech via a pass called the Tizi-n-Test (2,092m) and provides an excellent opportunity to explore wonderful mountainous landscapes as well as the birdlife of middle altitudes. The road will be described from the south (Souss valley) to the north (Marrakech area).

Location

The Tizi-n-Test road (S501) starts in the Plain of the Souss at its junction with the main P32 road. Marrakech lies 171 kilometres by road from this point, with 140km being in the mountains; the pass itself is reached after 37km.

Accommodation

Most birders travelling this road have either stayed at Marrakech or Taroudannt the previous night (see 'The Souss Valley' page 88) but several hotels are available along this road for anyone wishing to break the journey. Stopping en route also allows more time to explore mountains, and can be especially pleasant when it becomes hot in the plains on either side. There is no decent accommodation on the southern side of the pass (the 'Café Sunset' 8km before the summit was too dirty in 2001) and the first good choice is the basic but pleasant – though without heating – Hotel 'Bellevue' (❸ ☎ 061.38.76.22 or 067.05.58.44) near the pass. On the northern slopes, the only hotels are at Ouirgane; the good 'Sanglier qui fume' (❷ ☎ 044.48.57.07 and 08, Fax 044.48.57.09), the old and nice but expensive 'La Roseraie' (❶ ☎ 044.43.91.28 and 29, Fax 044.43.91.30), and a bed-and-breakfast 'Chez Momo' (❸ ☎ 044.48.57.04). Some 4kms later, on the left side of the road, there is the newly built, wonderful 'Bergerie' (❶ ☎ 044.48.57.16 and 17, Fax 044.40.57.18) which is probably the best choice if you can afford it; it is wise to book well in advance during school holidays. There was no hotel available at Asni in 2001 (the old 'Toubkal' was closed) and nothing beyond here until Marrakech, which has a full range of accommodation. However, accommodation is available in Imlil (see page 65).

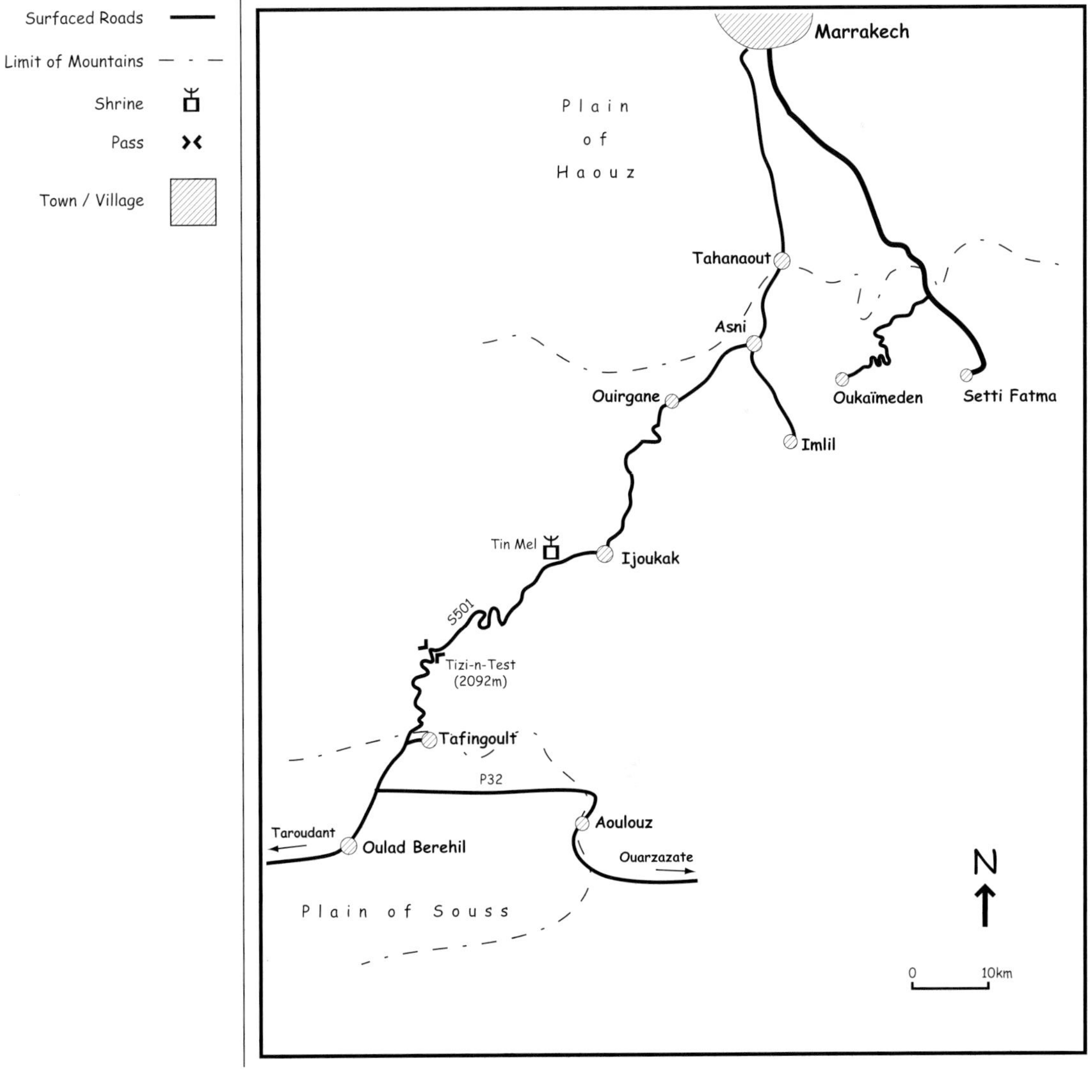

Strategy

The journey is wonderful in springtime, but less so in summer when temperatures can be very high. The pass is sometimes blocked by snow in winter (December-March) and the road is then impassable between +22km and +56km from the junction with the P32 (barriers will prevent you going higher than these points); during this period, it is wise to ask advice from the local authorities before leaving Taroudannt. In spring and summer, beware of sunburn at high altitudes (for example if dining on the terrace of the Hotel Bellevue).

Plan a full day for the journey. The first eight or so kilometres after the P32 junction run straight across the Argan forest and are typical of the Plain of the Souss; the mountain road then begins with its many curves and hairpins all the way up to the summit and you will find many opportunities to stop en route for gorgeous scenery and birds. In springtime, two recommended stops are the little waterfalls 3 or 4km before the summit, and the Hotel Bellevue near the summit for the scenery, the visible passages of birds, and the great 'Berber Omelettes' which can be enjoyed on the sunny terrace.

The northern slopes are greener as they receive more rain; after a while the road runs along the Oued N'Fiss and it is then worth spending an hour or so exploring the riparian woods. Stop to visit the Tin-Mel Mosque which was built in the 12th century and is one of the very few that non-Muslims can visit in Morocco.

Birds

The argan forest near Tafingoult is good for species such as Hoopoe, Rufous Bush Robin, Sardinian and Orphean Warblers, Moussier's Redstart, Woodchat and Southern Grey Shrikes, Black-crowned Tchagra, and Magpie (Moroccan race). Turtle Doves and finches (Serin, and the local races of Chaffinch, Greenfinch, Goldfinch and Linnet) are among the commonest birds.

Moussier's Redstart

In the Aleppo pine reforestation and scrubs likely species include Turtle Dove, Common Cuckoo, Rufous Bush Robin, Blackbird, Black Wheatear, Tristram's Warbler, Great Tit, Chaffinch, and the local race of Crossbill (*poliogyna*). Common Bulbul occurs up to +21km at least;

Moussier's Redstart can be found everywhere, and Rock Bunting is fairly common at middle elevation.

Red-billed Choughs and wild Rock Doves breed in the chalk cliffs (concretions) that overhang the road near the summit. Blue Rock Thrush, Black Wheatear, Kestrel, and Peregrine (some years) breed in the cliff near the Hotel Bellevue.

Birds of the Holm oak coppices and the Phoenician and Spanish juniper woods include Sparrowhawk, Barbary Partridge, Woodpigeon, Moussier's Redstart, Tristram's, Subalpine, Sardinian and Western Bonelli's Warblers, Serin, and the local races of Mistle Thrush, Firecrest, Coal Tit, Blue Tit, Great Tit, Short-toed Treecreeper, Jay, Chaffinch, Linnet, and Hawfinch.

The riparian woods along the Oued N'Fiss are the habitat for Turtle Dove, Common Bulbul, Wren, Robin, Nightingale, Blackbird, Cetti's, Olivaceous and Melodious Warblers, Blackcap, Spotted Flycatcher, Blue and Great Tits, Golden Oriole, Chaffinch, and Serin.

In Asni there is a Cattle Egret and White Stork colony in the trees at the Gendarmerie Royale. Shortly after Asni the road runs along the Oued Reraya (Grey Wagtail, Wren, and Cetti's Warbler) and then goes through the Moulay Brahim Gorges (Bonelli's Eagle in the 1980s); Little Swifts and Red-rumped Swallows breed under the bridge at the exit of the Gorges.

Other Wildlife

The area is good for reptiles and some Bibron's Agamas will probably be visible sunbathing near the road; local Moroccan Day Geckos live on rocks (for example at the small waterfalls near the pass on the southern side). Evidence from road kills has confirmed that Moorish Viper occurs in the area.

Imlil

(6b – Western High Atlas)

Imlil is a small, typical village of the Western High Atlas accessed through one of the nicest valleys of the region. It is also the starting point for the climb to the Jbel Toubkal (the highest peak in North Africa, 4,167m) and one of the only regular places where White-rumped Swift can be seen in Morocco.

Location

The village lies some 60km south of Marrakech, at the end of a 17km dead-end road starting from Asni (see 'The Tizi-n-Test Road' page 62). This small, winding road can be in bad condition especially after snow or heavy rainfall and slow, careful driving is essential.

Accommodation

There are two hotels at Imlil. If the price is of no concern, try the fantastic but expensive 'Kasbah du Toubkal' perched above the village with a stunning view of the Toubkal (❶ www.kasbahdutoubkal.com ☎ 044.48.56.11, Fax 044.48.56.36). This was one of the authors' favourite hotels in Morocco in 2002. Otherwise, the medium-category 'Etoile du

Toubkal' is good value for money (❷ / ❸ ☎ 044.43.43.87 and 44.97.67, Fax 044.43.43.87).

The Club Alpin Français (☎ 022.27.00.90, Fax 022.29.72.92) offers basic dormitory accommodation in its refuge; another (better) solution is to stay in one of the few 'Gîtes d'étapes' (bed-and-breakfasts) available, e.g. 'Chez Jean Pierre' (☎ 022.48.56.09 and 044.44.91.05). Alternatively, accommodation is available in the N'Fiss valley (see 'The Tizi-n-Test Road' page 62) or in Marrakech.

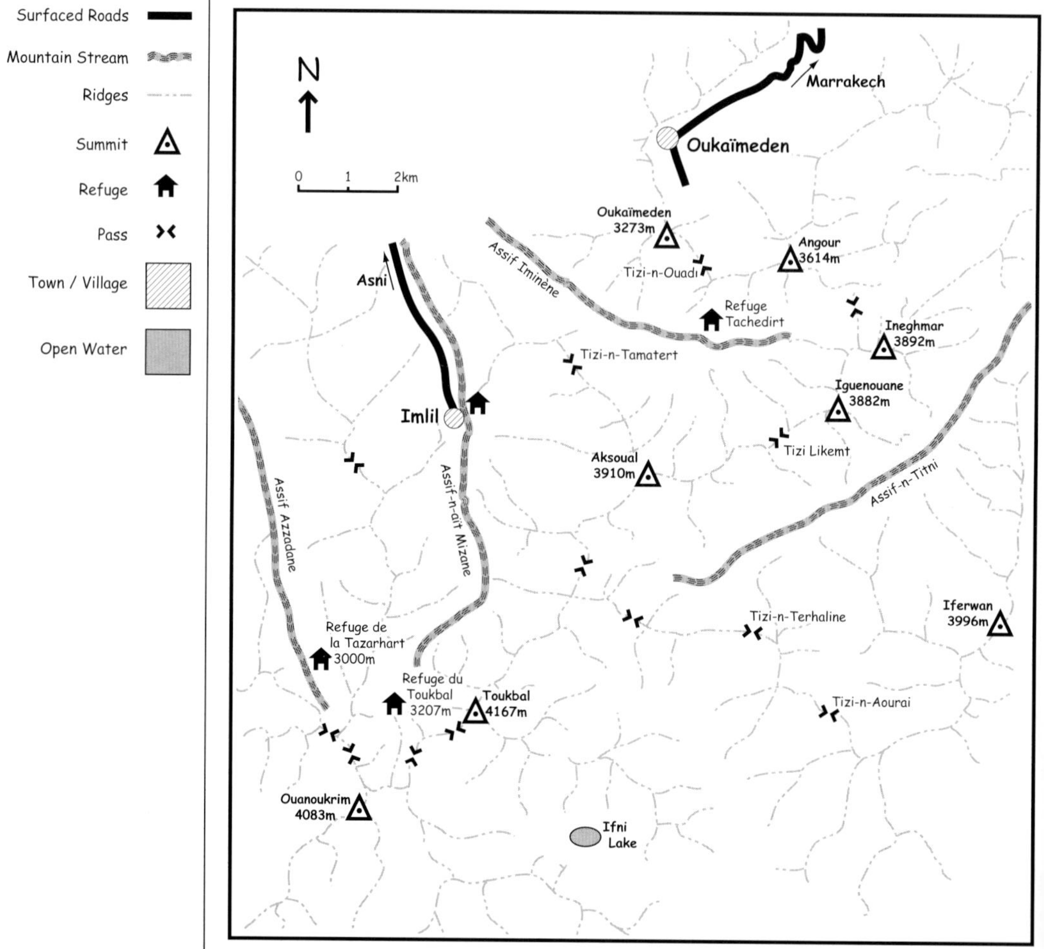

Strategy

The valley is under snow in winter and thus is best explored from March/April until November (although winter trekking is developing quickly). Bear in mind that the White-rumped Swifts don't arrive until late May and depart in September.

Imlil can be visited as a day-trip from Marrakech but staying overnight is recommended, especially in late spring and summer when it becomes uncomfortably hot in the plains and in the Atlas foothills. The greenness of the valley, the numerous brooks, and the shelter of the large walnuts are then very welcoming and relaxing. Cars should be parked in the village – from there, walk along one of the numerous footpaths. Carry food (available in the village) if intending to take a long walk.

This is the perfect starting point for those planning to climb the Jbel Toubkal. Specialised high-mountain guides can be found at the 'Bureau des Guides' in the village, or can be contacted through the hotels listed above. The climb and return takes two days with a night at the 'Refuge du Toubkal' (3,207m).

Birds

The authors have recently described the avifauna of the Marrakech region, including the Asni valley and Imlil in a paper in the French journal *Alauda*, so those requiring detailed information on the birds of this region should refer to that (see bibliography). Copies are available from the authors.

Two very sought-after birds can be seen at Imlil. White-rumped Swifts were discovered here in mid-July 1969, when some 30 birds were found probably breeding in the cliffs between the village and the Refuge du Toubkal. Since then, they have been regularly recorded in this area and in the nearby valleys but always in small numbers. They arrives in late May (earliest record: 18 May 1978) and leave by early October (latest record: 3 October 1979). Their habitat is mainly cliffs, but perhaps also villages. Very little is known about their biology and the authors welcome any reports. The Levaillant's Woodpecker is another Moroccan speciality, which can be found in the walnut trees of the valley, such as those just under the 'Kasbah du Toubkal' for example.

Other interesting high-altitude birds often recorded here include Red-billed Chough and Alpine Chough, which can be seen in flocks of several hundred birds. Local people try hard in springtime to prevent them from eating all their cherries.

The riparian woods are good for species such as Turtle Dove, Common Bulbul, Wren, Robin, Nightingale, Blackbird, Cetti's Warbler, Blackcap, Blue and Great Tits, Golden Oriole, Chaffinch, and Serin. Pied Flycatcher is a common passage migrant. Atlas Flycatcher is very scarce and local in the High Atlas but was found here, feeding its young at a nest in a Great Spotted Woodpecker hole in June 1985. Grey Wagtail inhabits the river beds and Rock Bunting is not uncommon on the nearby slopes.

Oukaïmeden
(6b – Western High Atlas)

This high altitude site in the Toubkal massif of the High Atlas ranges from 2,600m, where a skiing resort has been built, to 3,600m at Jbel Angour. About 100 species have been found here including some of the rarest passerines in the country. At least 50 species breed. For detailed information refer to Barreau, D.; Bergier, P. & Lesne, L. 1987. L'avifaune de l'Oukaïmeden, 2,200-3,600m (Haut Atlas, Maroc). *Oiseau et R.F.O.* 57: 307-367. Copies are available from the authors.

Location

Oukaïmeden is 72km south of Marrakech. The only easy way to get there is by car, as it cannot be reached by bus. Traffic is light except on winter weekends for skiing. Follow the S513 from Marrakech towards the Atlas to the Vallée de l'Ourika. Shortly after reaching the mountains take the 6035A, a good tarmac road on the right-hand side which leads to Oukaïmeden.

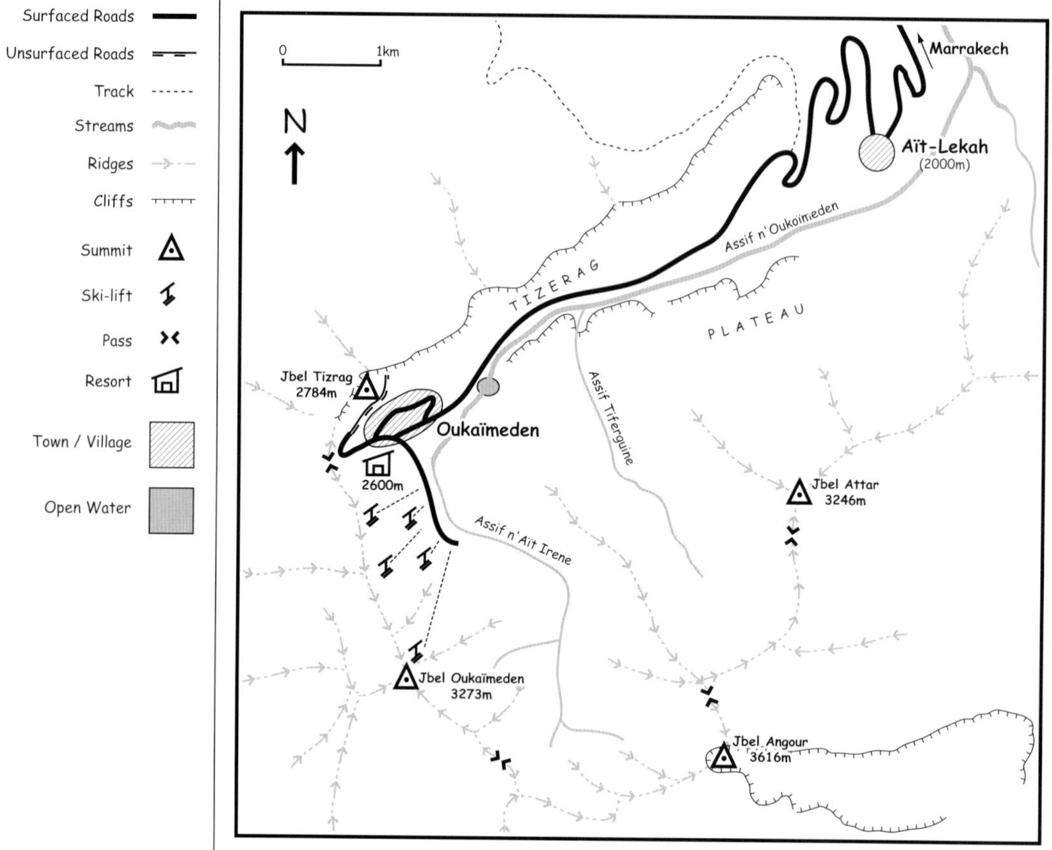

Accommodation

Some hotels and restaurants have been built at the resort: the 'Kenzi' (❶ ☎ 044.31.90.80 to 86, Fax 044.31.90.88), the 'Imlil' (❷ ☎ 044.31.90.32; closed from late April to December), and 'L'Angour' (also called 'Chez Juju' ❷ ☎ 044.31.90.05, Fax 044.31.90.06. Good bird-log). Accommodation is also available at the chalet of Club Alpin Français (❸ ☎ 044.31.90.36, Fax 044.31.90.20) – attractively priced but bedding is not always supplied, especially in the summer. From spring to autumn, it is possible to camp near the caravans in the meadow in front of the resort. The above-mentioned hotels also act as restaurants, but it is a good idea to bring a picnic lunch. There are few other basic restaurants which serve the usual tajines. Make sure your car has enough petrol for the return journey as there are no petrol stations here.

Strategy

The drive from Marrakech takes about 90 minutes so it is possible to visit the site from there as a day-trip or even half-a-day if you start early. It is worth staying a couple of days at this site in late spring and summer when it is uncomfortably hot in the plains.

The end of spring, summer, and autumn are the best seasons to visit the site. Temperatures then are reasonably high (20-25°C by day and not too cold at night). The sun is almost always shining and care should be taken not to get sunburned.

The drive from Marrakech through the plain of Haouz passes through various habitats; arid with Jujube trees *Ziziphus lotus* up to 900m, semi-arid with Sandarac trees (900-1,200m), temperate with Holm Oak (1,200-1,900m), cold semi-arid with Thuriferous Juniper (1,900-2,500m), and finally the high mountain layer with low thorny bushes above 2,500m. Each habitat has its own bird community and regular stops should enable a good cross-section of species to be seen.

Winter is cold, but during sunny days birds come to feed close to the ski station and nice views can be had of Crimson-winged Finches, Shore Larks, Rock Sparrows, Alpine Choughs, and Red-billed Choughs.

Birds

On the drive from Marrakech look for: Rufous Bush Robin on the plains; Little Swift, Red-rumped Swallow, and the Moroccan race of White Wagtail, mainly in the foothills; Moussier's Redstart, Tristram's, Spectacled and Subalpine Warblers in junipers and *Genista*; and Black Wheatear and Blue Rock Thrush wherever there are rocks; this latter can be fairly common, e.g. 60 were counted along the 6035A on 3 October 1999. The very rare White-rumped Swift has been seen near Imlil in the Asni valley, 20km to the west (see that site page 65).

The resort is at the lower limit of the alpine layer. On the plateau to the east there are Alpine Swifts, Tawny Pipits, Black Redstarts, Black Wheatears, Blue Rock Thrushes, Alpine and Red-billed Choughs. One of the commonest passerines is the Moroccan race of Northern Wheatear and Horned Lark is also abundant, especially on the slopes at the edges of the plateau. The poorly known Crimson-winged Finch can also be seen here and small flocks often occur. Lammergeier was regularly seen in the mid-1980s but has much decreased after widespread poisoning campaigns in the whole of Morocco.

In the village Rock Sparrows are common having colonised the buildings in the early 1950s. A walk towards the summits of the Jbels can be good for seeing Alpine Accentor and Rock Thrush. Sunset from the top of Jbel Tizerag, easily accessible by a track, is an unforgettable sight.

Other wildlife

Various reptiles hide amongst the rocks; North African Eyed and Iberian Wall Lizards, and especially the High Atlas Moroccan Day Gecko, are all common. Barbary Ground-squirrel is hard to miss. The rock engravings on the plateau at the east of the resort are worth visiting. Look for the Robust Marsh Orchids all along the 'assifs' (streams) in spring.

The Tizi-n-Tichka road

(6b – Western High Atlas)

This is the most important road that crosses the central part of the High Atlas and is the route most used when travelling between the 'northern regions' (Marrakech) and the pre-desert (Ouarzazate – page 113). It reaches its highest point at a pass called the Tizi-n-Tichka (2,260m).

Location

The distance by road between the towns of Marrakech and Ouarzazate is 204km and the pass is almost half way.

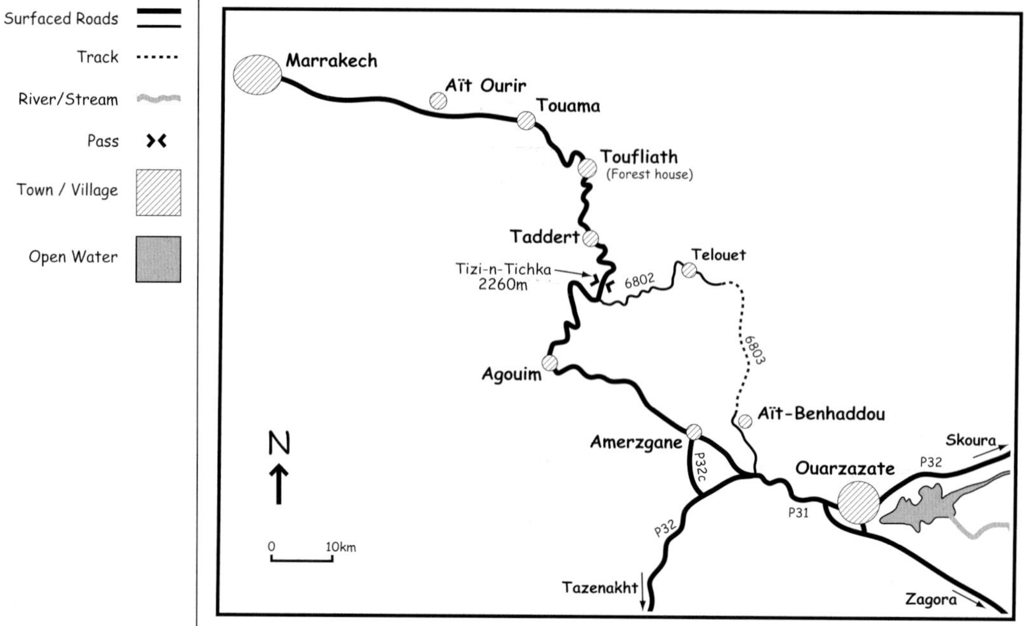

Accommodation

There is little decent accommodation available en route between Marrakech and Ouarzazate except 'le Coq Hardi' at Aït Ourir, 38km from Marrakech (❸ ☎ and Fax 044.48.00.56) and 'Dar Oudar' at Touama, 38km from Marrakech (❸ ☎ 044.48.47.72), both excellent value for money. However, there are many small restaurants in most villages, e.g. at Taddert on the northern slopes and at Agouim on the southern side. Hotels in Ouarzazate are listed in the 'Ouarzazate and the Barrage Mansour Eddahbi' section, page 113.

Strategy

It takes about four hours to drive the 204 kilometres without stops, but it is best to plan a full day for the journey, to allow sufficient time to enjoy the birdlife of the High Atlas.

The pass can be blocked by snow in winter despite every effort made by the snow ploughs which generally operate quickly to keep it open. Also be careful when driving as there are often many lorries and coaches that go very fast.

Good stops en route on the northern side include the forest house of Toufliath for the various local races of passerines, Great Spotted and Levaillant's Woodpeckers, and the few tall trees 12km to the south of Taddert – in front of Café Atlas Tizi-n-Tichka – for Crossbill (*poliogyna*) and other passerines. If time allows, visit the Glaoui Kasbah at Telouet, a vast complex of Berber buildings abandoned in the mid 1950s. In 2002, the tarmac road 6802 (21km from the main P31 to Telouet through wonderful forests of Phoenician and Thuriferous junipers) was in good condition. However, the track 6803 from Telouet to Aït-Benhaddou (some 40km) was impassable with a regular vehicle and was best tackled with a four-wheel drive (or, better still, by mountain bike or on foot!).

Birds The authors have recently described the avifauna of the Marrakech region, including the Tichka area in a paper in the French journal *Alauda*, so those requiring detailed information on the birds of this region should refer to that (see bibliography). Copies are available from the authors.

The northern slopes are covered with Holm oak and Phoenician and Spanish junipers. Likely species include Sparrowhawk, Booted Eagle, Peregrine, Barbary Partridge, Woodpigeon, Moussier's Redstart, Tristram's, Subalpine, Sardinian and Western Bonelli's Warblers, Serin, and the local races of Mistle Thrush (*deichleiri*), Firecrest (*balearicus*), Coal Tit (*atlas*), Blue Tit (*ultramarinus*, a good candidate to become a full species), Great Tit (*excelsus*), Short-toed Treecreeper (*mauritanica*), Jay (*minor*), Chaffinch (*africana*), Linnet (*mediterranea*), and Hawfinch (*buvryi*). One of the most sought-after birds in Morocco, Levaillant's Green Woodpecker, is fairly common here and you should search for it near the forest house of Toufliath. Crossbills (*poliogyna*) breed in the Aleppo pine plantations and Goshawk is a likely breeder.

The riparian vegetation along the rivers are the habitat of Common Bulbul (up to 1,800m), Wren (3,200m), Rufous Bush Robin (1,500m), Robin (1,800m), Nightingale (1,800m), Blackbird (2,600m), Cetti's Warbler (1,900m), Western Olivaceous and Melodious Warblers (2,200m), Spotted Flycatcher (1,800m), and Golden Oriole (1,700m).

Shore Lark, Tawny Pipit, Alpine Accentor, Northern Wheatear (*seebohmi*), Crimson-winged Finch, Rock Sparrow, Alpine and Red-billed Choughs have been recorded in the alpine-like vegetation near the pass, but they are less frequent than at Oukaïmeden (page 67).

The cliffs are good for Kestrel, Peregrine and Blue Rock Thrush (the Rock Thrush is much rarer). Lammergeier has almost disappeared after large poisoning campaigns and every record should be reported to the GOMAC; Barbary Falcon has been positively identified at the pass.

The southern slopes are drier with mainly steppe-type vegetation with some scattered junipers or other bushes. Interesting species include Black-eared and Black Wheatears, and Rock Bunting.

Crag Martin is an uncommon breeder but is very common in winter on both northern and southern slopes; African Rock Martin has been claimed but there is no definite proof of its presence here.

The typical birdlife of the pre-desert zone appears to the south of Amerzgane. For example, the regs (stone desert) and stony hills along the P32c are good for Desert Lark, Mourning Wheatear, and Trumpeter Finch.

The valleys and gorges of the Oueds Dadès and Todra

(6c – Central High Atlas)

These two rivers rise in the central part of the High Atlas. They flow south passing through impressive gorges and irrigate lush, green valleys. The Dadès gorge has many local fortified villages (kasbahs), and the Todra valley is lined with one of the most scenic palm-groves in the country. This beautiful landscape is worth visiting in itself and is a good example of the habitats to be found in the valleys of the southern foothills of the High Atlas. The more adventurous birdwatcher might consider driving from one valley to the other via Tamtattouchte. The desert immediately to the south of the P32, between the two valleys, is one of the most accessible and one of the best places to see a range of desert birds (see 'The Tagdilt track' page 103).

Location

Access to the valleys starts from the towns of Boumalne du Dadès and Tinerhir respectively, both located on the main P32 road that runs along the south side of the High Atlas. The two towns can be reached by bus but a car is really necessary to explore the area. The gorges can be reached by taxi but the prices can be exorbitant for tourists. From Boumalne du Dadès the S6901 runs up the Dadès gorge (tarmac up to Msemrir in 2002) and from Tinerhir the S6902 leads to the Todra gorge (tarmac up to the gorge in 2002, but extension is likely).

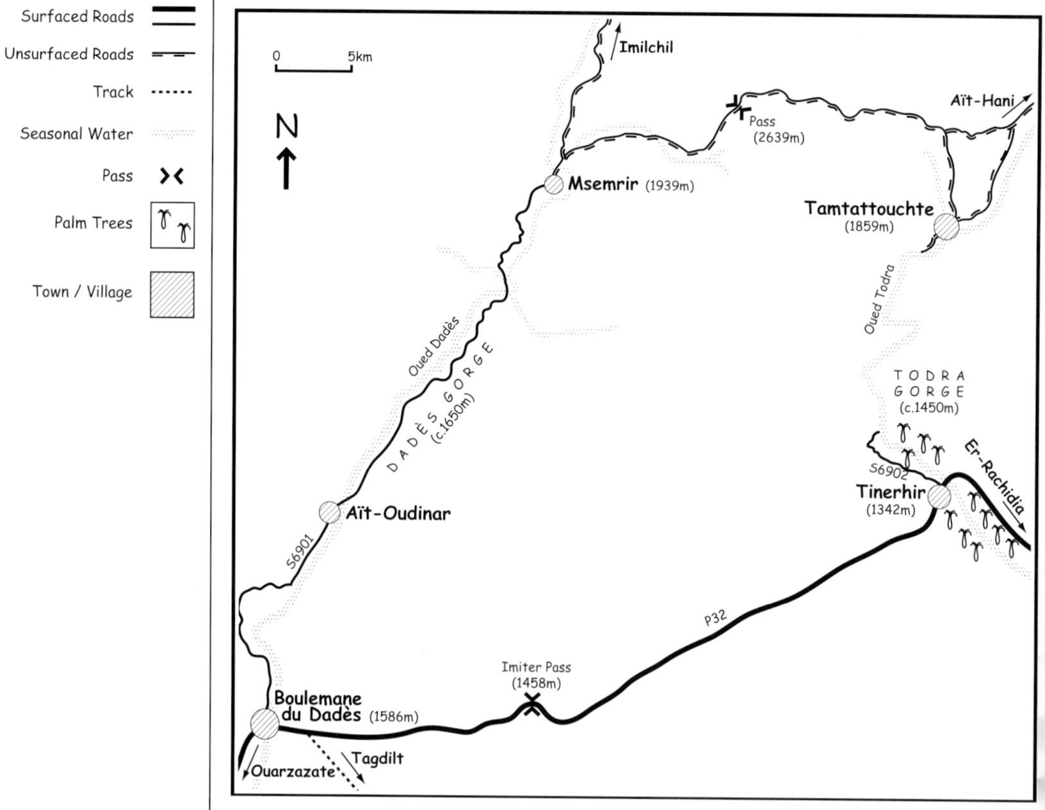

Accommodation in the Dadès Valley

Staying overnight in the Dadès valley is a very relaxing and pleasant experience after the heat of the desert. Several small, simple hotels and inns have been built along the gorge, including 'Hotel Kasbah' 15km from Boumalne, 'Chez Pierre', 'La Fibule du Dadès', 'Peuplier', 'Gazelle du Dadès', 'Kasbah de la Vallée' and 'Atlas Berbère' around 27-28km, 'Entre les Gorges' at 31km, and 'Taghia' and 'Kasbah des roches' at 33km. However, in 2002 the authors preferred the 'Auberge des Gorges du Dadès' at 25km (❸ ☎ and Fax 044.83.17.10) and 'Berbère de la Montagne' at 34km (❷/❸ ☎ 044.83.02.28). Several of these hotels also welcome campers.

For those intending to visit the Tagdilt area in the same day, an alternative is to sleep at Boumalne du Dadès (see 'The Tagdilt track' page 103).

Accommodation in the Todra Valley

Tinerhir has some hotels such as 'l'Oasis' or the 'Avenir' in the budget category, the 'Tombouctou' in the medium category, and the 4-star 'Kenzi Bougafer'. But here again it is pleasant to spend a night in the cheap or medium price camping-hotels in the valley or in the gorge. The 'Soleil' and 'Azlag' are at 8km from Tinerhir, the 'Lac Auberge', 'Les Poissons sacrés' and 'Amazigh' at 9km, the 'Mansour' and 'Etoile des Gorges' just before the entrance to the gorge. There are two small hotels inside the gorge itself – both set in stunning landscapes – the 'Yasmina' (❷/❸ ☎ 044.83.42.07) and the 'des roches' (❸ ☎ 044.83.48.14, Fax 044.83.36.11). These were the authors' favourites in 2002, even though the standard left a little bit to be desired.

Strategy

Winters can be very cold here and the best time to visit the valleys is from March to May and the mountains are best from May to July. It is worth spending at least a day here. Half a day should be spent in the palm-groves and in the Todra gorge and the other half exploring the Dadès valley. Walking on the paths along the wadis amidst palm, olive, almond, fig or pomegranate trees and passing the stunning kasbahs and through villages built of mud is an unforgettable experience.

If you intend to drive the circuit between the two valleys through the Atlas, start at Tinerhir, going through Tamtattouchte and Msemrir to Boumalne du Dadès. A four-wheel drive and a guide were essential in 2002, as the tracks were ill-marked and in poor condition.

The region, and especially the city of El Kelaa des Mgouna some dozen kilometres to the west of Boumalne du Dadès, is famous for its roses, which bloom in April-May. It is possible to buy rose-water to scent your linen and dry roses which will scent your home for more than a year.

Birds

In the palm-groves, orchards and cultivated fields of the lower valleys White Stork, Hoopoe, Turtle Dove, Scops Owl, Common Bulbul, Nightingale, Blackbird, Cetti's and Western Olivaceous Warblers, Blue Tit, Golden Oriole, Serin, and Goldfinch all breed. Great Spotted Woodpeckers breed in the Dadès valley.

At the gorges Long-legged Buzzard, Golden and Bonelli's Eagles, and Lanner are sometimes seen; Lammergeier has almost disappeared after large poisoning campaigns. Rocks and cliffs are the habitat of Rock Dove, House and Crag Martins (Todra gorge), Black Redstart, Black Wheatear, Blue Rock Thrush, and Rock and House Buntings. Grey Wagtail breeds near water.

Going further north into the mountains, the rocky plateaux, screes, and scrub vegetation contain such species as Barbary Partridge, Desert Lark (up to c.2,200m, despite its name), Moussier's Redstart, Scrub,

Spectacled and Tristram's Warblers, Red-billed Chough, Trumpeter Finch and, depending on the date, perhaps the last of the wintering Ring Ouzels. One of the most interesting species here is probably Crimson-winged Finch (which also sometimes comes in to drink in the Todra gorge).

Above 2,600m Horned Lark and the North African race of Northern Wheatear (*seebohmi*) can be found among the high altitude xerophytic vegetation.

The two valleys are excellent for migrants, which follow these routes to cross the High Atlas. Common conspicuous spring migrants include European Bee-eater, Barn and Red-rumped Swallows, House Martin, Redstart and Pied Flycatcher, but many others pass through, including some waders, such as Common and Green Sandpipers.

The Hassan-Addakhil Dam and the Ziz valley near Er-Rachidia

(6d – Eastern High Atlas)

This is one of the two major reservoirs in the south, set in the wonderful desert landscape of the Eastern part of the High Atlas. It is often worth a visit when travelling through the region as interesting species have been recorded here. If you have spare time, the palm-groves along the Oued Ziz downstream from the barrage, from Meski southward, are also worth visiting.

Location

Er-Rachidia is located at the strategic crossroads of the north-south P21 road from the Middle Atlas to Tafilalt, and the west-east P32 road that runs from the Atlantic to the Eastern Sahara. The Hassan-Addakhi Dam was built in 1971 on the Oued Ziz to the north of the town, at the outlet of the Eastern High Atlas; the Oued Ziz is normally perennial throughout the year in its upstream (Atlas) part and feeds the reservoi with an average flow of six cubic metres per second, but this flow can be much reduced in dry years, such as in the early 2000s. Because of the dam, the Oued is generally dry downstream until it reaches Meski where the abundant flow of the 'Source bleue' spring waters flood it again down to Erfoud and Rissani.

Accommodation

Er-Rachidia is a large but not very touristic town; nevertheless it has a range of hotels including the 'Ansar' (❸ 34 rue Ibn Batouta. ☎ 055.57.39.19), 'Renaissance' (❸ 19 rue Moulay Youssef. ☎ 055.57.26.3 and 'Royal' (❸ 8 rue Mohamed Zerktouni. ☎ 055.57.30.68) in the budget category; the 'M'Dagha' (❸/❷ 92 rue M'Dagha. ☎ 055.57.40.4 48 and 49, Fax 055.79.08.64) is recommended as although it is a bit mo expensive it is good value for money. The 'Kenzi Rissani' (❶ Avenue Moulay Ali Chérif. ☎ 055.57.25.84 and 21.86, Fax 055.57.25.85) is a 4-star, and the most expensive hotel of the town. It is also worth considering staying at Erfoud (see page 105) which is only 80km to the south.

There is a campsite with a 'natural' swimming-pool at Source bleue de Meski, a nice palm-grove 21km to the south of Er-Rachidia on the P21 road to Erfoud; it was poorly maintained at the end of the 1990s but it has now improved.

Many restaurants can be found in Er-Rachidia, especially in the main street (Avenue Moulay Ali Chérif), but no food is available near the reservoir so it is best to take a picnic.

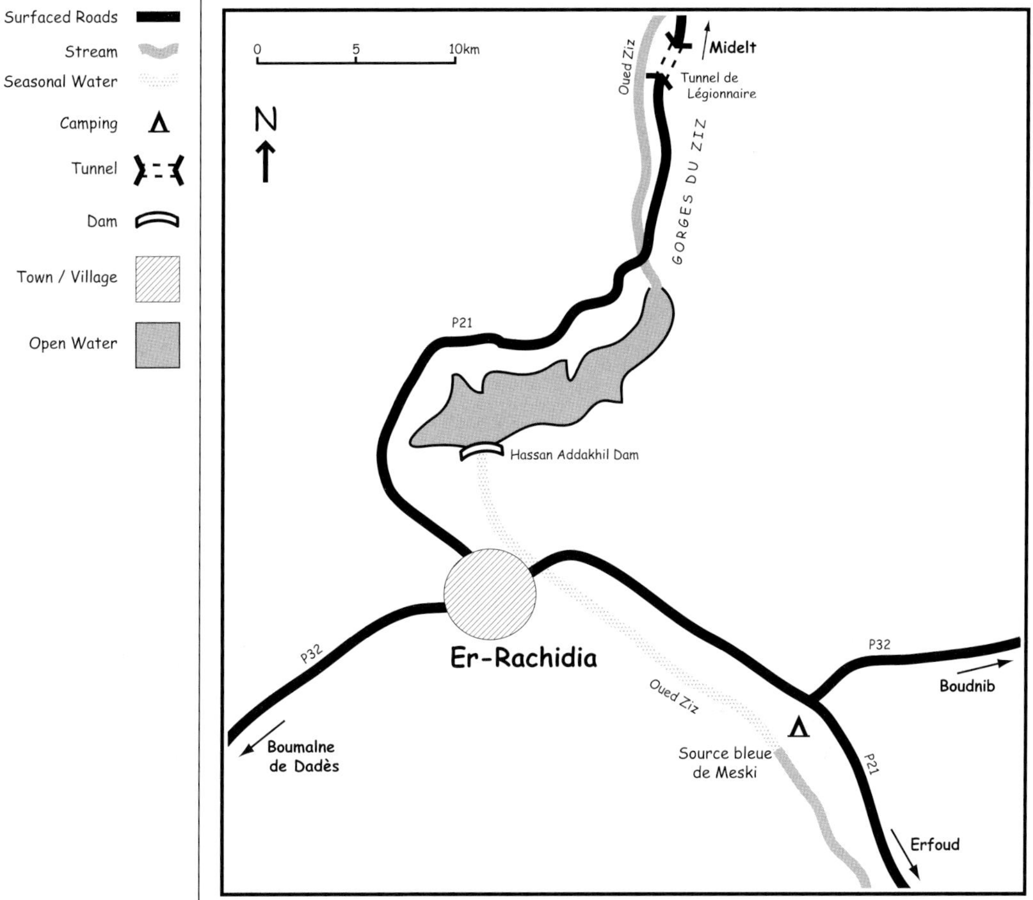

Strategy

The migration periods, especially in springtime from March until May, are the best times to visit the area; summers are very hot and birdwatching is then difficult and less productive; winters can be good but are cold: it is sometimes freezing in the palm-groves.

The reservoir can be visited in a couple of hours and thus it can be a convenient mid-day halt when travelling from the Tafilalt (page 105) or the Upper valley of the Oued Moulouya (page 77) to the valleys and gorges of the Oueds Dadès and Todra (page 71). On the other hand, you should plan to stay at least a full day if you intend to explore the palm-groves of the Oued Ziz.

The access to the dam itself is restricted. The main P21 road which heads to the north (to Rich and Midelt) passes fairly far from the reservoir but there are two rough tracks that will get you closer to the water; you will find the first one at the milestone 'Rich 53' and the second one 3km later (milestone 'Rich 50'). A telescope is recommended.

The palm-groves along the Oued Ziz, between Meski and Erfoud, are also excellent for migrating birds and the best way to explore them is on foot, just with binoculars. You will probably quickly be accompanied by some friendly children; talk to them and show them some birds, but please don't give them anything as this encourages begging (remember the Tourist Code, see page 12).

Birds

Waterbirds at the reservoir include Great Crested Grebe, Grey Heron, Little Egret, Eurasian Coot, and several wintering or migrating ducks such as Pintail (mid-February through April), Wigeon (December–February), Common Teal (October-March), Northern Shoveler (October-April), and Pochard (October-May). This is a good place for Ruddy Shelduck, which breeds here with up to 200+ in April 2001. Black-necked and Little Grebes, Squacco Heron, Cattle Egret, Marbled Teal (up to 200 in January 1998), Common Shelduck and Tufted Duck are rarer. There are two records of Ring-necked Duck: a pair from 11 December 1980 to 1 February 1981 and a single bird on 25 January 1998; a strange place for a North American species – 500km from the Atlantic! A Great Egret (16 March 2001) was another Moroccan rarity.

The waders are generally not numerous as there are no extensive mudflats but Greater Flamingos and Spoonbills sometimes stop to refuel. Interesting spring migrants include Gull-billed and Black Terns, and Collared Pratincole. The European race of Great Cormorant (*sinensis*) is now a regular winter visitor.

The Ziz corridor is followed by many migrant raptors including Honey-buzzard, Osprey, Black Kite, Short-toed and Booted Eagles, Egyptian Vulture, Marsh and Montagu's Harriers, and Hobby. Migrations of passerines such as Common Swift and European Bee-eater, may also be impressive.

The rocky, desert slopes around the lake are the habitat of Bonelli's Eagle, Long-legged Buzzard, Lanner, Peregrine (African race *minor*), Barbary Falcon, Kestrel, Barbary Partridge, 'Pharaoh' Eagle Owl (*ascalaphus*), Thekla, Desert and Thick-billed Larks, Crag Martin, Desert Mourning, Black and Black-crowned Wheatears, Blue Rock Thrush, Spectacled Warbler, Southern Grey Shrike, and Trumpeter Finch. Large groups of Common Ravens have been recorded (up to 400 on 22 April 1982).

The palm-groves and nearby cultivated areas downstream from Meski are good for a large variety of migrating waterbirds including Water Rail, Spotted, Little and Baillon's Crakes, Moorhen and Little Ringed Plover (which also breeds), Little Stint, Redshank, Spotted Redshank, Greenshank, Wood and Common Sandpipers, and Jack Snipe. Green Sandpiper and Common Snipe are both winter visitors and passage migrants.

Other migrants include Common Cuckoo, European and Red-necked Nightjars, Roller, Hoopoe, Wryneck, Nightingale, Bluethroat, Northern and Black-eared Wheatears, Redstart, Whinchat, Rock Thrush, many warblers (among which Reed, Sedge, and Subalpine Warblers are the commonest in both spring and autumn), Pied Flycatcher, Woodchat Shrike, Golden Oriole, and Ortolan Bunting. Isabelline Wheatear is probably a rare but regular spring passage migrant.

In winter Stonechat, Black Redstart, Blackbird, Chiffchaff, Blackcap and Sardinian Warbler are common. Tristram's Warbler is fairly common in the oases (and especially in the tamarisks near Merzouga, see page 106) and Dartford Warbler rarer. Ring Ouzel, Starling, Spotless Starling, and Siskin are among the less frequent winter visitors.

Breeding birds of the palm-groves and gardens along the Ziz valley include Turtle, Laughing and Collared Doves, Long-eared, Tawny and Little Owls, Kingfisher, Common Bulbul, Cetti's and Western Olivaceous Warblers, Rufous Bush Robin, Blackbird, Fulvous Babbler, Spotted Flycatcher, Great and Blue (North African races) Tits, Serin, Greenfinch, and Goldfinch. Blue-cheeked Bee-eater occurs as far north as Meski along the Ziz valley.

The Upper Valley of the Oued Moulouya and Zeïda

(7c – High Moulouya)

The Oued Moulouya is the second largest oued in Morocco. Its upper course crosses vast areas of semi-desert covered with Halfa grass (*Stipa tenacissima*) and Wormwood bushes (*Artemisia herba-alba*), lying between 900 and 1,500m in elevation. This unique region's special habitats contain interesting passerines, among which are several wheatears and larks, including the much sought-after Dupont's Lark.

Location The most interesting parts of this region, which lies just to the southeast of the eastern Middle Atlas and to the north of the High Atlas, are within an area bordered by the towns of Imouzzer-des-Marmoucha, Boulemane, Zeïda, Midelt, and Missour.

Accommodation The only decent hotels in the region are in Midelt, the 'El Ayachi' being the most expensive (❷ Rue Agadir. ☎ 055.58.21.61). Several other basic hotels are very good value for money, and the 'Atlas', run by a Berber family (❸ 3 rue Mohamed Amraoui. ☎ 055.58.29.38), and the 'Boughafar' (❸ 7 Avenue Mohammed V. ☎ 055.58.30.99) are recommended. There is a camp site ('Timnay') some 10km to the south of Zeïda, but there is no problem camping in the open countryside in this area. Cafés and basic restaurants are common in every town, especially at Zeïda, which is an important crossroad, and at Midelt (those near the bus station serve succulent 'tajines', and there is a tiny restaurant at the Hotel Atlas).

Strategy There is very little traffic on any of the roads in the region except the

P20 and P21, making it difficult to hitch-hike and so it is almost impossible to birdwatch here without a car. Winters are very cold and long on these high plateaux, summer and autumn are generally very hot and less interesting for the birdwatcher. The best time to birdwatch here is from April through to June. In May, it is often possible to birdwatch all day long, but in June the afternoons can sometimes be hot. Stay out until dusk to listen for 'Pharaoh' Eagle Owl and Red-necked Nightjar.

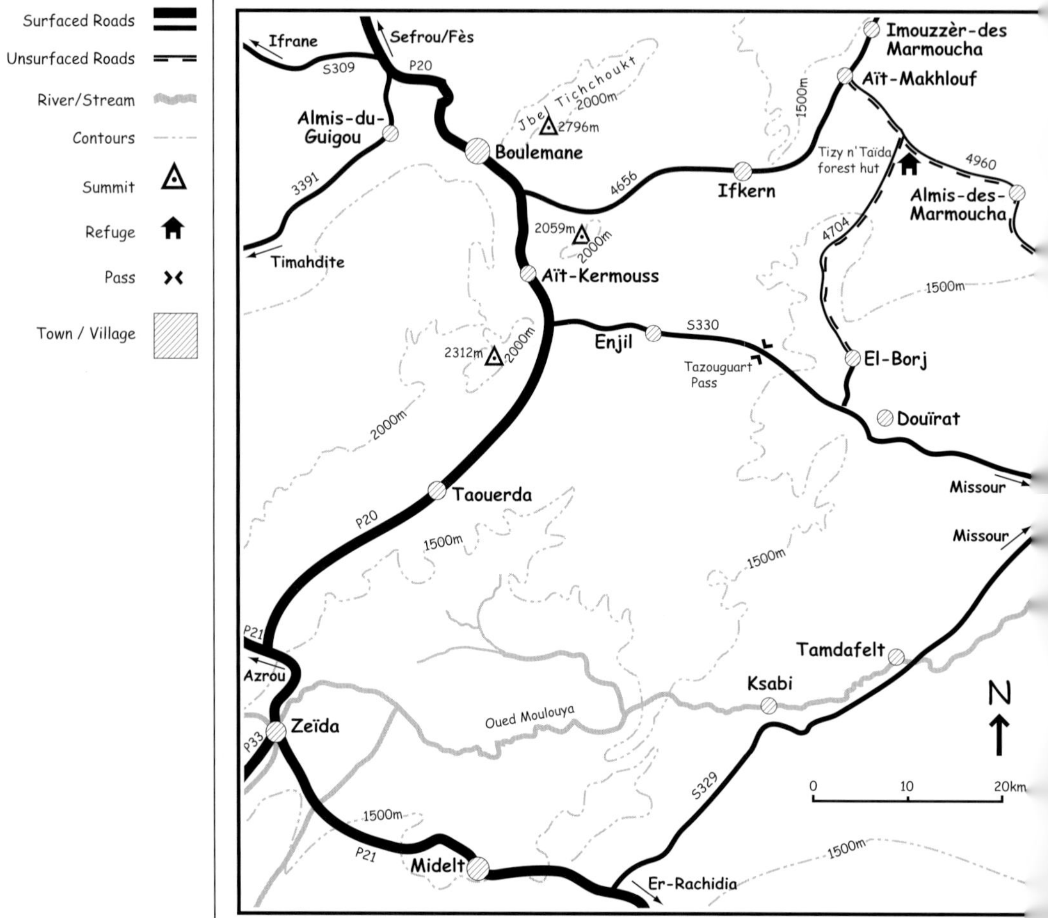

One of the best ways to appreciate the ecological complexity of this area is to drive slowly and observe the distribution and density of birds, especially larks and wheatears. It is worth walking in the Halfa steppe to experience this special habitat. If time is available it is worth exploring the mountain tracks to the northeast of Imouzzer-des-Marmoucha where Booted Eagle, Scops Owl, and Subalpine and Orphean Warblers can be seen. The mountains Jbels Bou Iblane and Bou Naceur, the highest points of the Middle Atlas located to the northeast of the area, have been studied by few birdwatchers and may be worth a visit.

Because of limited time, most birders only go to Zeïda to look for Dupont's Lark, and this area will be described in a separate section which follows this one.

Birds

The distribution of birds in this area is governed by altitude, rainfall and vegetation. Larks and wheatears are probably the most fascinating groups of species to observe here.

On the plateaux of the Middle Atlas (along the S309 coming from Ifrane) Skylark, Horned Lark, Tawny Pipit, Northern Wheatear (Moroccan race *seebohmi*), Black-eared Wheatear, and Spotless Starling can be seen.

Several kilometres south of Boulemane the P20 passes through semi-arid and then arid habitats. Follow the P20 or turn east onto the S330. Desert and Temminck's Horned Larks, and Desert and Red-rumped Wheatears can be found here while there are no Horned Larks any more (because of a lower altitude), and Skylarks and Black Wheatears are rare. The Moroccan race of Northern Wheatear only rarely occurs in this semi-arid habitat and is confined to a few particular places such as around the village of Almis-des-Marmoucha, while the European races are a common occurrence during migration.

The transitional area between the slopes of the Middle Atlas and the plateaux of the upper Moulouya, where the two avifaunas meet, are particularly interesting. In a small field 5km north of Taouerda, Crested/Thekla, Calandra and Temminck's Horned Larks, and Skylarks have all been seen together in May.

Besides the already mentioned birds, several other interesting species may be seen in the region. The cliffs and rocks attract raptors such as Black Kite, Long-legged Buzzard, Booted Eagle, Kestrel, and Lanner, while Egyptian Vulture has dramatically suffered from large poisoning campaigns. Other species found here include 'Pharaoh' Eagle Owl, Crag Martin, Blue Rock Thrush, and Rock Sparrow.

Birds of the Halfa steppe include Barbary Partridge, Stone-curlew, Black-bellied Sandgrouse, Red-necked Nightjar, Short-toed Lark, and Tawny Pipit. Trumpeter Finches are often present in good numbers.

The gardens and adjoining areas near the Oued Moulouya (such as at Missour or Ksabi) hold various breeding birds among which are Turtle Dove, European Bee-eater, Rufous Bush Robin, Nightingale, Western Olivaceous and Orphean Warblers, Spotted Flycatcher, and Southern Grey Shrike.

In the mountainous areas to the north of Immouzèr-des-Marmoucha birds such as Hobby, Scops and Little Owls, European Nightjar, Redstart and Moussier's Redstart, Melodious, Orphean and Western Bonelli's Warblers, Woodchat Shrike, and Cirl and Rock Buntings can be seen.

During the spring the Moulouya valley acts as a migration corridor for several species including Osprey, European Bee-eater, and Subalpine and Melodious Warblers.

Other Wildlife

Several mammals live in the Halfa steppe. Greater Egyptian Jerboa is common and often seen dead on the road. Fox, Jackal and Algerian Hedgehog may also be seen and Leopard has even been reported in the 1980s from the holm-oak forest near Imouzzer-des-Marmoucha. Beware of scorpions which hide under the stones.

Zeïda

(7c – High Moulouya)

Zeïda is the best area to look for one of the most poorly known Moroccan birds, Dupont's Lark. (See 'The Upper valley of the Oued Moulouya and Zeïda' (page 77) for Location and Accommodation.)

Strategy

In spring time, Dupont's Larks are particularly vocal very early in the morning and stop singing shortly after dawn – sometimes sooner. Consequently early morning visits or overnight camping is essential, but the region can be very cold and it is wise to bring warm clothes – at 6.00 am on 21 April we have recorded a temperature of 9°C, with a strong wind and a little rain, and no further songs of the lark were heard after that time. Songs have also been heard in autumn (e.g. at 5.30 am on 25 September) and in late winter.

Despite considerable effort some birders have missed this bird, as it is one of the oddest species in Morocco. The probability of hearing it i much higher than seeing it, and a tape recorder for playback can be very useful.

To actually see the bird, one of the best strategies is to walk across t steppe while looking far ahead; the bird often runs away fast through the tussocks, and hides quickly. Another strategy is to put a playing tape recorder on a bush and watch from a car nearby, waiting for a bi to come and look for the 'intruder'.

The plains some kilometres to the south of Zeïda are one of the traditional places to search for the bird. From the bridge over the Oue Moulouya at Zeïda, tracks will be found at 1.7km (to the southwest o the P21), at 3.5km (on both sides), at 4.0 and 5.5km (to the southwest), and then at 10.2km just after the Oued Anseguenir on both sides of th road. Follow any of these tracks for a few hundred metres to enter suitable habitat, though basically the whole steppic region is good for the species (for example on 21 April 2001 in addition to several singin birds on the track at 3.5km – approximately 500 metres southwest fro the road – two others were heard from the main road just in front of the huge hill midway between Zeïda and Midelt).

Birds

Dupont's Lark is widely distributed throughout the Eastern Moroc High Plateaux, from Aïn Beni-Mathar in the north to the Figuig area in the south, and to Zeïda in the west. There are only a few records outside of this large area, mainly in the semi-desert south of the High Atlas. Halfa grass and wormwood steppes around 1,000 metres form its basic habitat. The authors only have details of two breeding recor (a nest nearly completed on 1 February, and adults feeding young on 7 June) so much is still to be discovered about this species.

Lesser Short-toed Lark is a resident breeder, while Short-toed Lark a migrant breeder, present from March to September/October. Thick-billed Lark is less common, and probably erratic. It favours light steppes, and sometimes feeds in ploughed or cereal fields; breeding takes place from mid-March through June. Thekla Lark is fairly common, as is Temminck's Horned Lark. The status of Crested Lark i uncertain. In winter, Shore Larks and Skylarks sometimes come dow from their Middle Atlas breeding ranges.

The area is also good for wheatears. Northern Wheatear is only a passage migrant, but the Moroccan race (*seebohmi*) breeds in the nearby Middle Atlas. Black-eared also passes through and breeds in the foothills. Desert is a migrant breeder present from mid-April to September/October, though some individuals overwinter. Red-rumped and Black Wheatears are both resident.

Black-bellied Sandgrouse is often seen here, but breeding has never been proved. Other steppic birds include Cream-coloured Courser, Stone-curlew, and Trumpeter Finch.

Spotless Starling is common at Zeïda. Collared Dove is a new and successful colonizer (first record in 1992) and Laughing Dove has been recently discovered here.

The National Park of Souss-Massa

(8 – Souss)

The National Park of Souss-Massa covers the whole coast of the Souss region and includes the estuaries of the two major oueds of that region, the Souss and the Massa. Its size varies from 65 to 70km in length and 10 to 15km in width; it covers 33,800 ha of land, and extends 1.5km out to sea. The boundaries are made up by the north bank of the Oued Souss to the north, and the village of Sidi Moussa d'Aglou to the south. A buffer area (42,200 ha) has been created to the east, up to the Aït Melloul/Tiznit main road (P30).

More than 270 species have been recorded so far in the park, including 90 breeding birds, including the famous Bald Ibis. The park also provides protection to some 40 mammal species, and more than 30 reptiles and amphibians.

The two most famous sites (the mouths of the Oueds Souss and Massa) are described below. An additional small, but interesting spot (the bridge over the Oued Massa) is not included within the limits of the park but is also described below.

Accommodation

At the north end of the park, Agadir is one of the most popular resorts along the Atlantic coast and has a wide range of hotels to suit all tastes; among the medium-priced, the authors can recommend the 'Hotel des Palmiers' (❷ Avenue du Prince Héritier Sidi Mohamed. ☎ 048.84.37.19, Fax 048.82.25.80) and the 'Hotel Petite Suède' (❸ Avenue Hassan II. ☎ 048.84.07.79 and 00.57, Fax 048.82.00.57). Inezgane is also a very convenient city, some kilometres south of Agadir and close to the mouth of the Oued Souss. Mid-range hotels here include 'La Pyramide' (❷ Chemin de l'Oued Souss. ☎ 048.83.47.05), 'La Pergola' (❷ Route d'Agadir. ☎ 048.83.08.41), and 'Le Provençal' (❷ Route d'Agadir. ☎ 048.83.26.12).

To the south of the park, the city of Tiznit also has many hotels, and good choices include the 'Hotel de Paris' (❸ Avenue Hassan II. ☎ 048.86.28.65, Fax 048.60.13.95) and the nearby 'Tiznit Hotel' (❷ ☎ 048.86.24.11 and 38.86, Fax 048.86.21.19).

The mouth of the Oued Souss
(8 – Souss)

The Oued Souss flows into the Atlantic just south of Agadir. The mouth of the river is very wide with mud and sandbanks; it is include in the National Park of Souss-Massa and this is a strategic resting plac for numerous waders, gulls, and terns. The site has become one of the most famous birdwatching places in Morocco.

Location The site lies just to the south of Agadir, and is reached by taking the main road southeast from town, to Tiznit and Taroudannt. At Inezgane, which is now part of the suburbs of Agadir, follow the right-hand side road to the mouth ('Embouchure du Souss'), close to the 'Golf du Soleil' and 'Golf des Dunes'. The Royal Palace is anothe good landmark.

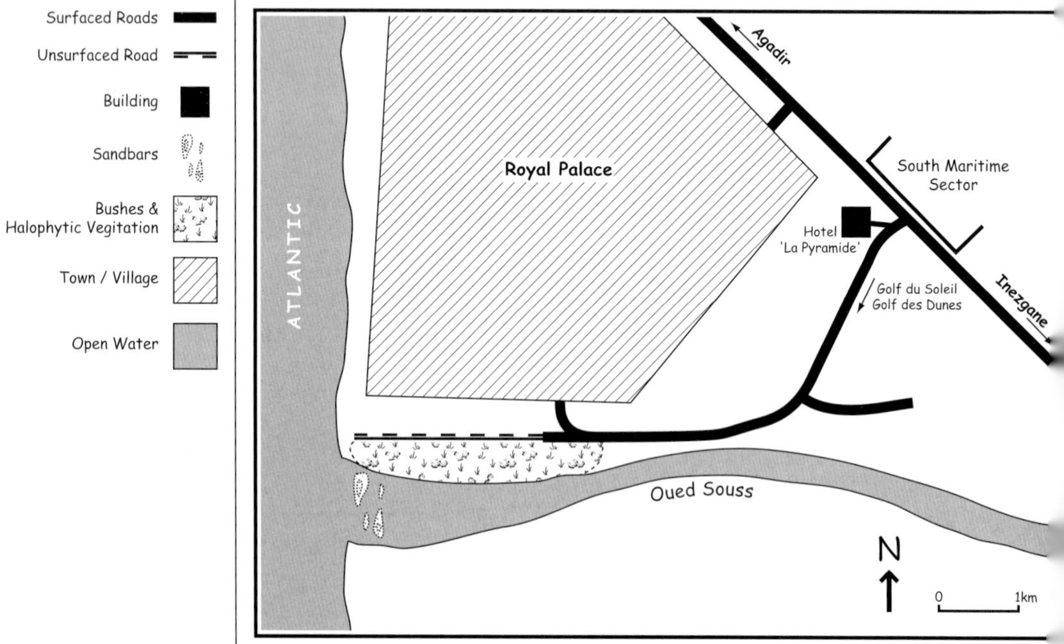

Accommodation No accommodation is available at the site itself, and it is most convenient to stay in Agadir or Inezgane (see above).

Strategy The migration periods of February-April and September-Novembe are the best times to visit, but the winter can also be interesting, especially at the end of the year when the first 'spring' migrants appe Summer is probably the least interesting time. The site is better at hig tide. Remember to lock cars when parking them, and make sure any valuables are out of sight.

Some observation platforms and numerous paths have been built along the northern shore of the Oued. Follow these paths towards the ocean, but avoid walking in the halophytic vegetation or on the mudflats, as this might disturb the birds. A telescope is very useful at this site. A few hours are generally sufficient to see most of the birds present, but for those based nearby it is worth returning several time as the turnover of birds is high. A visit at night is also worthwhile for species such as Stone-curlew, Long-eared Owl, and Red-necked Nightjar.

Birds

The site is famous for its waterbirds, waders, gulls, and terns, and for the impressive list of rare species that have been recorded.

Little Grebe breeds, but Great Crested and Black-necked are only winter visitors. Among the regular large waterbirds are Great Cormorant (*maroccanus* race), Little Egret, Grey Heron, White Stork, Spoonbill, and Greater Flamingo. Purple Heron and Glossy Ibis are less common. Bald Ibis sometimes visit the site to feed. The site is not very suitable for ducks, but Ruddy Shelduck is sometimes seen.

Waders are often numerous in migration periods and in winter, and the commonest include Oystercatcher, Black-winged Stilt, Avocet, Little Ringed, Ringed and Kentish Plovers, Grey Plover, Red Knot, Sanderling, Little Stint, Curlew Sandpiper, Dunlin, Ruff, Black-tailed and Bar-tailed Godwits, Curlew, Redshank, Spotted Redshank, Greenshank, Green, Wood and Common Sandpipers, and Turnstone. There are several records of Marsh Sandpiper and just a few of Dotterel. Northern Lapwing is only seen during harsh winters. Stone-curlew and Barbary Partridge breed in the dunes.

The mouth is very good for gulls and terns, which use the sand bars as roosting places. Black-headed, Lesser Black-backed and Yellow-legged Gulls, and Sandwich Tern are common, and it is not unusual to find one or more of the following species: Mediterranean, Little, Sabine's, Audouin's, Common, and Great Black-backed Gulls, Kittiwake, and Gull-billed, Caspian, Royal, Lesser Crested, Common, Arctic, Little, Whiskered, Black, and White-winged Black Terns, especially in winter and spring. This is one of the best Moroccan sites for Slender-billed Gull.

This large concentration of waterbirds attracts raptors, and Marsh Harrier, Bonelli's Eagle, Lanner, and Peregrine and Barbary Falcons are sometimes seen hunting over the estuary.

The nearby bushes and wooded areas are excellent places to look for migrant passerines. Interesting breeding species include Plain Martin, Yellow and Moroccan White (*subpersonata*) Wagtails, Common Bulbul, Zitting Cisticola, Western Olivaceous and Sardinian Warblers, Black-crowned Tchagra, Southern Grey Shrike, Moroccan Magpie, and Spotless Starling. House Bunting is common wherever there are houses, especially in Agadir. The eucalyptus woods are good for Woodpigeon and Blackbird, and numerous finches including Serin and Linnet. Booted Eagle, Long-eared Owl, and Red-necked Nightjar also breed here.

The list of uncommon or vagrant species which have been seen here includes Great Northern Diver, Bulwer's Petrel, Western Reef Egret, Great Egret, Lesser Flamingo, Spur-winged Goose, American Golden Plover, Great Knot, Semipalmated, Pectoral, Broad-billed, Buff-breasted, Terek and Spotted Sandpipers, Long-billed Dowitcher, Lesser Yellowlegs, Laughing, Franklin's, Bonaparte's, Grey-headed, and Ring-billed Gulls, African Skimmer, Richard's Pipit, and Red-eyed Vireo.

The mouth of the Oued Massa
(8 – Souss)

The Oued Massa flows from southeast to northwest across the Souss Plain reaching the Atlantic between Agadir and Tiznit. It is the most northerly Saharan wadi in Morocco. A sand bar often blocks the mouth of the river creating a lake which is then connected with the ocean only at the highest tides. Towering sand dunes run along the southern edge of this lake while the northern shore is fringed with mud banks and flat areas covered with glassworts and other halophytic vegetation.

Because of the beautiful landscape and the rich fauna this is one of the most famous birdwatching sites in Morocco. It has been a biological reserve since 1980, and is part of the National Park of Souss-Massa.

Location

The site is located on the Atlantic coast 40km south of Agadir. Because it is some distance away from any main roads or towns the easiest way to get there is by car. From Agadir follow the main P30 road south towards Tiznit; there are two small roads to the right that go towards the oued, the first one at km 57 (signposted 'Massa – Sidi-Rbat'), the other one some 3km further south, near a petrol station 'Petrom' (signposted 'Massa'). Once in the villages, turn right and driv north. Shortly after the village of Arhbalou the tarmac stops; the track that skirts the oued leads to the warden's house (and the biological reserve entrance). Park here and walk along the wadi, downstream to the mouth.

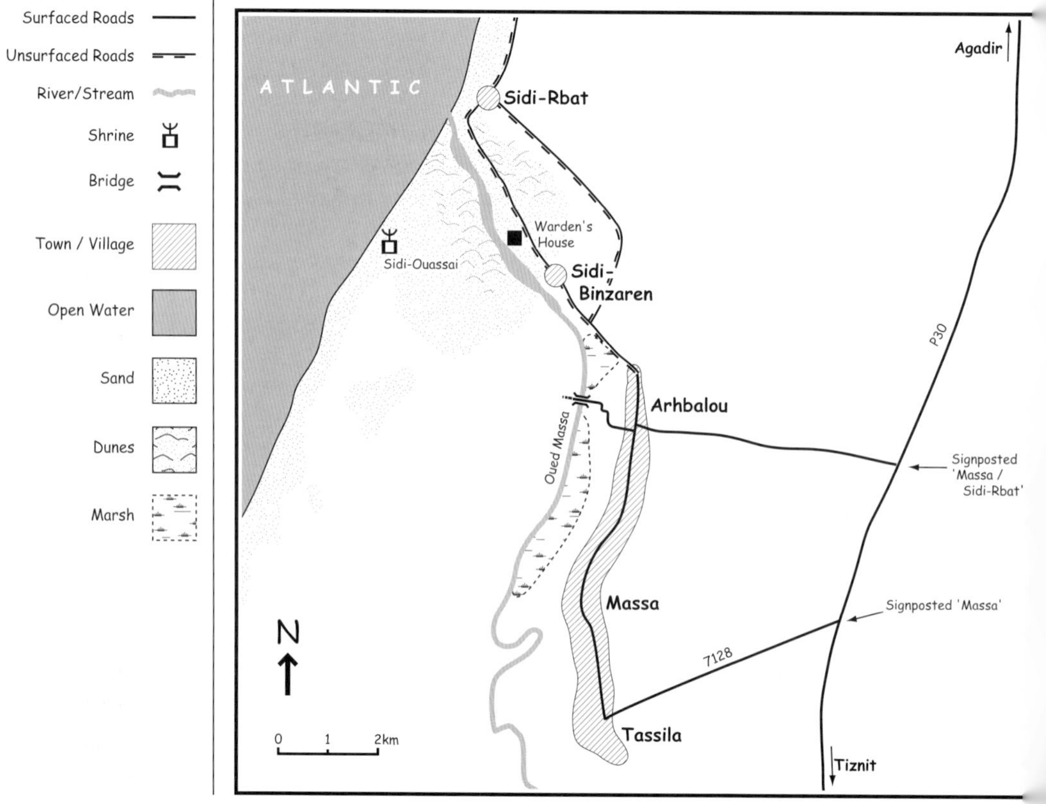

Accommodation

The campsite at the mouth of the oued (Sidi-Rbat) has been turned into a pleasant but very expensive hotel (❶ ☎ 022.94.02.59, Fax 022.94.02.22); a much cheaper alternative is to sleep at the 'Café-restaurant Le Musée' (❸ ☎ 048.26.02.10, GSM 062.41.82.62) at Sidi Abbou on the main road P30. Most birders stay in Agadir or Inezgane (see page 81). There is no food available, so bring a picnic or buy something in one of the villages upstream.

Strategy

The site is best visited as a day-trip from Agadir though its ornithological richness often deserves greater attention.

The best times to visit the site are between September and April when many winter visitors and migrants are present. A visit during March will produce both migrating and wintering species with some local birds also breeding. Summer is the least productive period. Post-breeding movements start in August but it is not until November that large numbers of birds are once again present at the site.

It is possible to birdwatch throughout the day, except during the hotter summer months. In the early morning it is worth sitting concealed near the mouth of the river to wait for Black-bellied Sandgrouse which come in to drink each day, approximately two hours after sunrise. Then walk along the track on the north side of the river but don't go right down to the shore as this might flush the feeding waders and waterfowl as well as annoy the warden. From November to early March Common Cranes fly in and spend the day in the reeds in the middle of the river across from the warden's house. They generally arrive between 10.00 and 11.00 and leave before sunset. Up to 200 birds have been counted in the winter.

If time permits, follow the track back to the point where the river turns south away from the track and look carefully in the reeds for species such as Little Bittern and Moustached Warbler. The few bridges over the oued upstream are also good vantage points; one of them is shown on the following sketch.

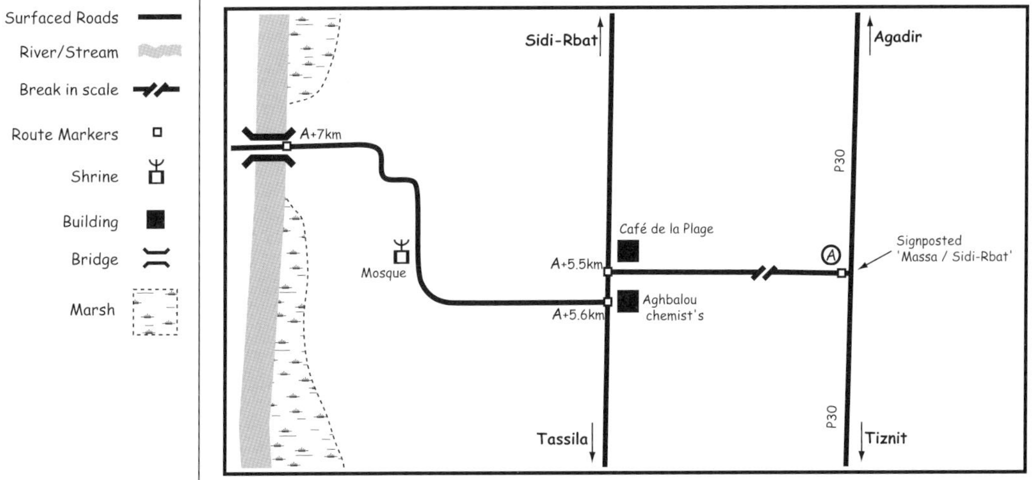

It is also worth walking across the desert areas to the northeast to look for Stone-curlews, Cream-coloured Coursers, and larks.

Seawatching may also be productive for petrels, shearwaters, gulls and terns. Do not bathe in the wadi as it is forbidden and there is a risk of disease. Be very careful if swimming in the ocean as there are strong currents close to shore.

Birds

During the winter the last five kilometres of the river hold an important population of Eurasian Coot (c.8,000) and numerous ducks (3,000-5,000). Most of the commoner European species are present as are some more unusual ones such as Ruddy Shelduck, Marbled Duck, Red-crested Pochard and Ferruginous Duck. The sand bar at the mouth of the river acts as a roosting place for gulls and terns (sometimes more than 2,000 birds). The commonest are Lesser Black-backed and Yellow-legged Gulls, but several other species are regularly seen including Audouin's and Little Gulls, and Caspian and Sandwich Terns; Royal and Lesser-crested Terns are rarer. The large number of birds at the site attracts raptors and Marsh Harrier, Black-shouldered Kite, Tawny, Booted and Bonelli's Eagles, and Peregrine and Barbary Falcons are all seen regularly.

The Richard's Pipit is among the rarest wintering passerines, and can be found in the cultivated fields. Plain Swift has been claimed several times here, but no certain proof is available.

The spring migration of some species starts as early as December. At that time the first Barn Swallows heading north sometimes meet the last Barn Swallows heading south! At Christmas the first Great Spotted Cuckoos, Pallid Swifts, Hoopoes, Red-rumped Swallows, House Martins and Reed Warblers appear. Migration continues until April. From February/March likely species include Little Bittern, Night, Squacco and Purple Herons, Glossy Ibis (occasional breeder), Spoonbill, Spotted and Baillon's Crakes, Gull-billed Tern, European Bee-eater, Bluethroat, Black-eared Wheatear, and various warblers, including Western Olivaceous, Spectacled and Subalpine.

For most species, breeding begins in March. Among the more interesting breeding species are Barbary Partridge, Stone-curlew, Cream-coloured Courser, Little Owl, Moussier's Redstart, Zitting Cisticola, Black-crowned Tchagra, Southern Grey Shrike, Spotless Starling, and House Bunting. The very distinctive local resident race of Great Cormorant (*P. c. maroccanus*, several dozen pairs) breeds from February to September; Plain Martin is fairly common and breeds throughout the year.

Three to six pairs of Marbled Ducks have bred regularly at the site since the beginning of the 1980s. Collared Dove colonised the area in the early 1990s, and is now a very common resident. Laughing Dove is fairly common in the palm-trees around the villages.

One of the most sought-after birds here is probably Bald Ibis, which breeds on nearby coastal cliffs. There were still 200-250 birds in 1998-9 including 60 breeding pairs. Please don't try to visit the birds at their colonies, as this could severely disturb the breeding success. They often come to feed in the Oued, and the last few hundred metres before the mouth is probably the best place to look for them.

Several species that do not actually breed at the site regularly visit it to roost or feed throughout the year. These include Little Egret, Greater Flamingo, Osprey, and Black-bellied Sandgrouse.

Many rare birds have been found at this site including some seldom seen elsewhere in the Western Palearctic. These have included Wilson's Storm-petrel, Western Reef Egret, Great Egret, Lesser Flamingo, Fulvous Whistling-Duck, Snow Goose, Spur-winged Goose, Black Duck, American Wigeon, Green-winged and Blue-winged Teals, Ring-necked Duck, Pallid Harrier, Allen's Gallinule, Long-billed Dowitcher, White-tailed Lapwing, and Ring-billed Gull. Keep a careful look out for such rarities.

Other wildlife

Several mammals are common here and these include Algerian Hedgehog, Brown Hare, Barbary Ground-Squirrel, Jackal, Red Fox, Weasel, Egyptian Mongoose, and African Wild Cat. Killer Whale has been seen offshore. Amphibians and reptiles are abundant and include Green Toad, North African Green Frog, Spanish Terrapin, Helmeted and Moorish Geckos, Bibron's Agama, Olivier's Small Lizard, Busack's Leopard and Golden Fringe-toed Lizards, Mionecton, Algerian Orange-tailed and Senegal Sand Skinks, Horseshoe, Hooded, Viperine and Montpellier Snakes, Schokar Sand Snake, and Moorish Viper. Egyptian Cobra is now very rare.

The bridge over the Oued Massa
(8 – Souss)

This is a tiny but often interesting spot which is worth a stop when travelling to/from the West Saharan coast.

Location

The area is located some 25km to the north of Tiznit, on the main road P30.

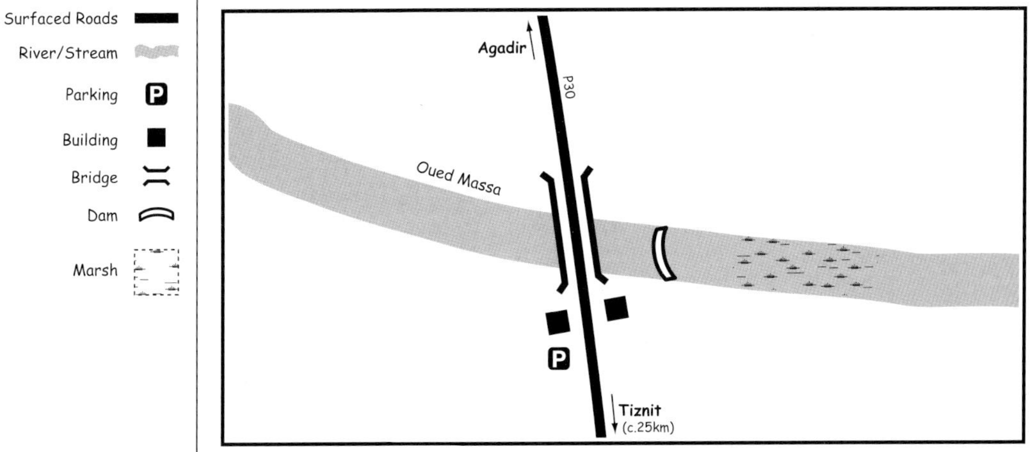

Accommodation

No accommodation or food is available here, and the closest hotels are at Tiznit.

Strategy

Driving between Agadir and Tiznit on the main road P30 is always laborious because of the volume of traffic and the coaches and lorries that travel (very) fast. It is a good idea to take a rest on this drive and this is the perfect place to stop for an hour or so.

Park under the eucalyptus trees on the western side of the road just to the south of the bridge, and go for a walk upstream (east) from the bridge to a small dam and a marshy area with tamarisks and reeds.

Birds

This is an excellent place for Little Bittern which is best looked for in April-May (a marked spring passage occurs in West Morocco from late March, occasionally earlier in February, and is over by early June; the autumn passage has rarely been reported, from August to mid-November). The species probably also breeds here in the reeds.

In spring, other likely species include Little Egret, Spoonbill, Black-crowned Night-Heron, Black-winged Stilt, Moorhen, Little Ringed Plover, Little, Pallid and Alpine Swifts, European Bee-eater, various species of swallows (including Red-rumped) and warblers, martins, White (Moroccan race *subpersonata*) and Yellow Wagtails, and Black-crowned Tchagra.

The nearby stone desert is good for Cream-coloured Courser, Tawny Pipit, Thekla Lark, Desert and Black-eared Wheatears.

Other Wildlife

Spanish Terrapin is common in the oued.

The plain of the Souss
(8 – Souss)

This plain, between the High and the Anti-Atlas, with its fruit orchards and vegetable cultivations is one of the richest areas of Morocco. Despite the expansion of this agriculture, the region still contains a very unique habitat, the argana forest, and is one of the most famous ornithological regions of the country.

Location

The region lies inland of Agadir and is bisected by the P32 which runs east from Aït Melloul to Ouarzazate. It can also be reached from Marrakech to the northeast along the S501 which crosses the High Atlas and climbs up to 2,092m at the Tizi-n-Test pass (see 'The Tizi-n-Test Road' page 62). The size of the region makes a car essential.

Accommodation

It is advisable to stay in Taroudannt, the historic capital of the Souss which lies at the centre of the region and has a number of hotels. The cheapest, low category hotels are downtown, the 'Taroudannt Hotel' being probably the best choice (❸ Place Al-Alaouyine. ☎ 048.85.24.16 but in this category the authors prefer the 'Hotel le Soleil' (❸ Dar Mbark, près Bab Targhount. ☎ 048.55.17.07), which is quieter being outside the walls of the city; two other excellent but more expensive choices are the 'Hotel Saadien' (❷ Bordj Oumansour. ☎ 048.85.25.89) and the 'Hotel Palais Salam' (❶ ☎ 048.85.21.30 or 23.12). If cost isn't a consideration try 'La Gazelle d'Or' (❶ ☎ 048.85.20.39 and 20.48), the most famous and expensive hotel in Morocco.

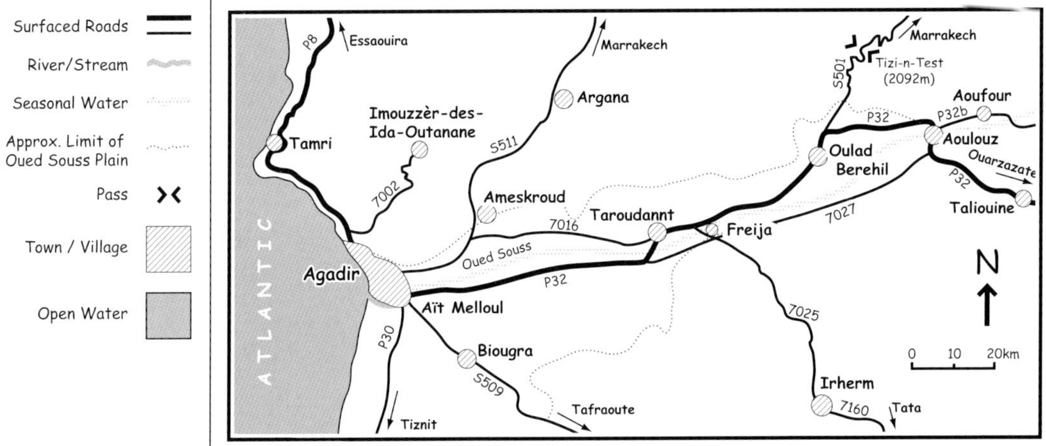

Strategy

The region contains a wide variety of habitats and several interesting or uncommon species. It is worth spending a full day here.

Like most sites, the Souss is at its best in the spring, but shouldn't be ignored at other times of year. It is possible to birdwatch throughout the day except during the hot summer months. During April and May it is worth staying in the argana forest until dusk to listen for Stone-curlew, Little Owl, and Red-necked Nightjar.

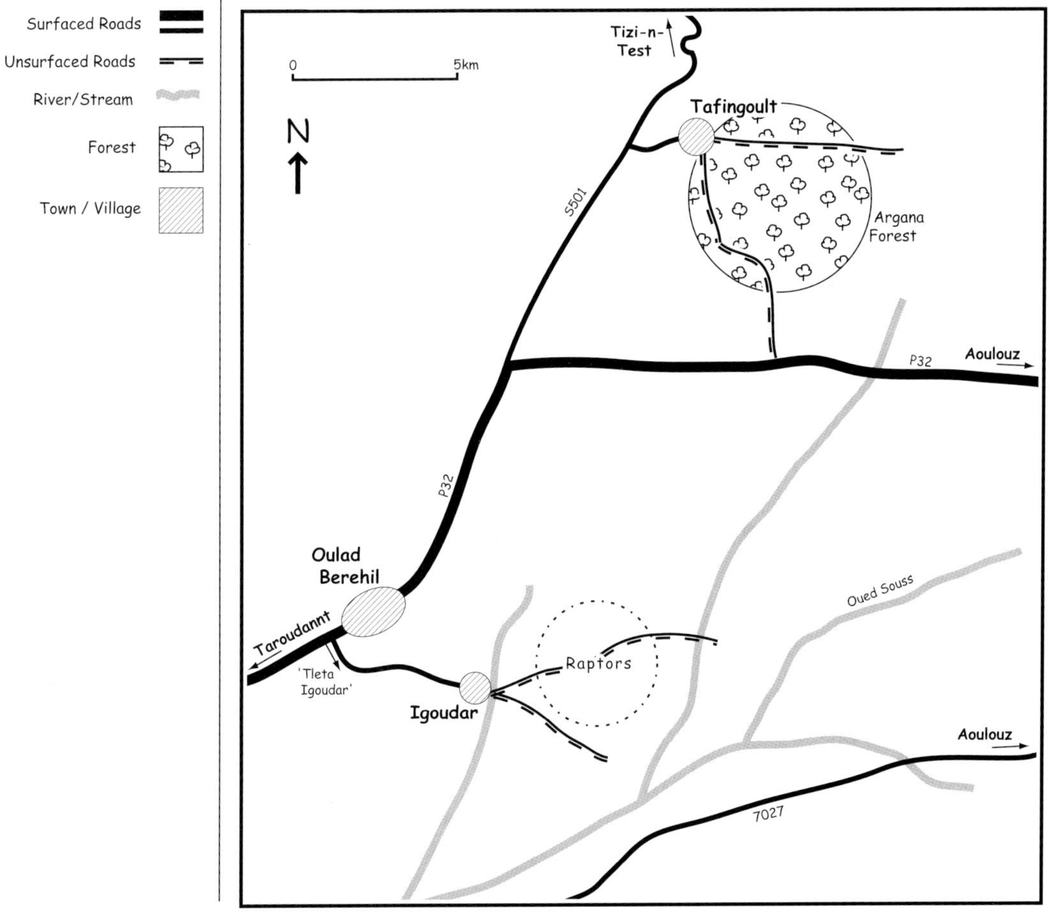

The argana forest is the habitat that generally attracts birdwatchers, and it is certainly one of the most interesting, but except in some remote areas, it has been degraded due to man or livestock; camels anc tree-climbing goats are a frequent sight beside the road in this area. Th best preserved patches of forest are at the north and the east of the region; the area near Tafingoult perhaps now being one of the best.

In the 1980s and 1990s, the most productive areas for raptors were near Freija (some 10 kilometres from Taroudannt on the road 7025 to Irherm) and at Igoudar (to the southeast of Oulad Berehil), but these sites have been less productive in recent years. The argana forest near Tafingoult is now an alternative choice.

Other areas such as the olive groves with their huge trees and the southwestern slopes of the High Atlas are all worth exploring. The are around the Oued Souss some 6km to the west of Taroudannt, where the P32 crosses the oued, has long been known as a good site (Black-shouldered Kite, Stone-curlew, Red-necked Nightjar, migrant passerines, Fulvous Babbler).

Birds

Many interesting species breed in the argana forest, including Stone curlew, Little Owl, Red-necked Nightjar, Hoopoe, Rufous Bush Robin, Sardinian and Orphean Warblers, Moussier's Redstart, Woodchat and Southern Grey Shrikes, Black-crowned Tchagra, and Magpie (Morocca race) whose nests are sometimes parasitised by Great Spotted Cuckoo. Turtle Doves and finches (Serin, and local races of Chaffinch, Greenfinch, Goldfinch, and Linnet) are among the commonest birds.

Black-shouldered Kite

The Souss is famous for raptors, particularly for three much sought after species: Black-shouldered Kite, Dark Chanting Goshawk, and Tawny Eagle. Black-shouldered Kite is certainly the most common of the three and up to eight have been seen in a single day; Tawny Eagle much rarer, and the last nests were found in argana trees in 1990 and 1992. Dark Chanting Goshawk is on the verge of extinction; several records were made in the area of Igoudar in the 1980s and early 1990s and the latest one (2002) is from 2000 when an adult was seen near Taroudannt on 15 April. In springtime, Black Kites, and Short-toed and Booted Eagles are common migrants. Long-legged Buzzard is fairly common in the north and the east of the plain, Bonelli's Eagle is rarer. Kestrel is the common falcon, but Lanner, Peregrine and Barbary Falcons are sometimes recorded.

One of the most famous breeding colonies of Bald Ibis was located in the east of the area near Aoulouz, in the Gorges of the Oued Souss, just upstream from the bridge. It disappeared when a dam was built in 1987, but the gorges are still good for many other species including Booted Eagle, Peregrine / Barbary Falcon, Black Wheatear, and Blue Rock Thrush. The reservoir in itself is not very good for birds (except for Ruddy Shelduck), but the landscape is very pleasant.

The arid areas along the road 7027 between Taroudannt and Aoulouz are good for Stone-curlew, Cream-coloured Courser, Black-bellied Sandgrouse, Crested / Thekla Larks, Black-eared Wheatear, Southern Grey Shrike, and Trumpeter Finch. Groups of Fulvous Babblers have also been reported.

The orchards and hedges contain lots of Turtle Doves and passerines, especially finches and they also form one of the favourite habitats of Black-shouldered Kite.

Old olive groves form a very pleasant habitat. Lots of birds can be seen here, especially during migration. Common breeding birds include Hoopoe, Common Bulbul, Blackbird, Rufous Bush Robin, Western Olivaceous Warbler, Spotted Flycatcher, Blue and Great Tits (Moroccan races), Spanish Sparrow, Cirl Bunting, and various finches.

Collared Dove (first record in 1987), Pallid and Little Swifts, Spotless Starling and House Bunting are all common in Taroudannt.

Fort Bou Jérif and the Oued Noun

(9a – Western Anti-Atlas)

This is a very spectacular area in the Western Anti-Atlas, essentially in the middle of nowhere. It is a nice place to stop and relax after several days (or weeks) of birding in the far south.

Location

The current 'Fort' Bou Jérif is a kind of tourist complex built in 1989 by a French couple, Guy and Evy Dreumont, near an old French Foreign Legion fort on the southern bank of the Oued Noun. Fort Bou Jérif is 40km from Goulimine; starting from there, follow the road north towards Sidi Ifni, but shortly after leaving Goulimine turn left to 'Laksabi – Plage Blanche'. 10km later, turn to the right to 'Laksabi – Plage Blanche' and then after another 3.5km turn right (signposted 'Fort Bou Jérif'). After 20km or so, the asphalted road turns into a good track; Fort Bou Jérif is after another 6km.

Accommodation

The very well kept tourist complex 'Bou Jérif Emirate'

(fortboujerif@yahoo.fr Maroc Aventure. B.P. 504. 81000 Goulimine, Fax 048.87.30.39) includes a motel (❸) containing five rooms with shared bathroom facilities – and a splendid hotel (The Caravanserail ❶) also with five rooms. There are also berber tents equipped for 5-6 people (❸ – campers and camper vans are also welcome. The restaurant at the complex has become renown over the years and among other specialities serves a famous camel tajine! Beer and wine are available. Advance booking is essential especially during the Easter period.

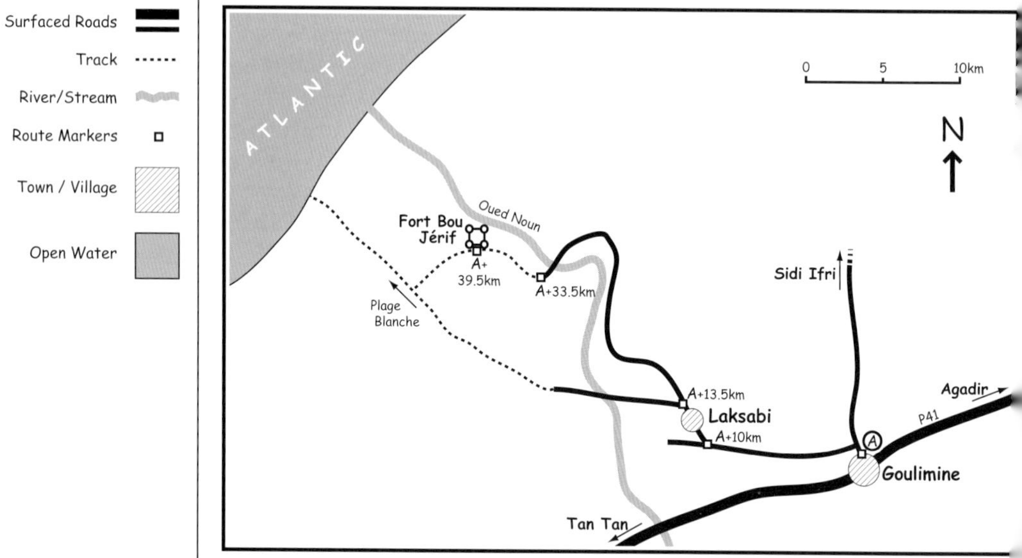

Strategy

After having spent some days or weeks birding in the areas further south without much comfort, we like to stay a night or two at this very relaxing place. The area also offers a good opportunity to explore the bushy steppe characterised by cactus-like euphorbias such as *Euphorbia echinus* and *beaumierana* (both spiny) and *E. rejis-jubae* (not spiny), mixed with *Senecio antheuphorbium*; this is a very special habitat specific to the Moroccan south-Atlantic coast.

Driving to the ocean and to the Oued Noun mouth on one of the many tracks often requires a four-wheel drive; the owner organises half-day or full-day excursions and can drive guests there. Starting from the tourist complex, it is also pleasant to walk along the oued to watch the many migrants that stop off here.

The road from Laksabi to Plage Blanche ('White Beach') was under construction in April 2001 and Plage Blanche is scheduled to become a tourist development area. This is a very beautiful long beach (some 40km between the Oueds Noun and Aoreora) and would be well worth visiting.

Birds

For some 30 kilometres after Goulimine, the road crosses vast plains covered with scant vegetation where Cream-coloured Courser, Black-bellied Sandgrouse, Desert and Red-rumped Wheatears, and Southern Grey Shrike live. 10km before Bou Jérif, the road overhangs the Oued Noun and this is a good place for Zitting Cisticola and Sardinian Warbler, Common Bulbul, Black Wheatear, House Bunting, and Black-crowned Tchagra. A Blue Rock Thrush was seen here among many other migrants in April 2000.

Shortly after crossing the oued, the tarmac road turns into a track; a pair of Bonelli's Eagles has built several nests on the first cliffs to the north which can be easily spotted with a telescope from the track (young were present in mid-April 2001). The euphorbia-covered steppe has a poor avian breeding community with very few species including Thekla Lark, and Black-eared and Red-rumped Wheatears.

Residents along the Oued include Barbary Partridge, Little Owl, Common Bulbul and Black Wheatear; many migrants stop en route and in April have included Little Ringed Plover, Common Sandpiper, large flocks of European Bee-eaters, swallows and martins, Yellow Wagtail, Redstart, Chiffchaff, Grasshopper, Subalpine and Orphean Warblers, Blackcap, and Golden Oriole. Lesser Whitethroat is a very rare migrant in Morocco and was recorded here on 5 May 1998.

Plage Blanche is an important wintering area for waders and Common Scoter. Other interesting species here have included Sooty Shearwater, Audouin's Gull and Lesser Crested Tern.

Other Wildlife

30 mammal species have been recorded in the area, including the endemic Tarfaya's Shrew, the Jackal and the Striped Hyena; Lesser Egyptian Jerboa is often found dead on the road near Laksabi; Barbary Ground-squirrel is common in the gorges of the Oued Noun. Spanish Terrapin is common in the oued. The region is also famous for its snakes, including the now very rare Egyptian Cobra and Puff Adder.

Taliouine

(9b – Central Anti-Atlas)

Taliouine is a very pleasant overnight stop in springtime, ideally located between Ouarzazate (page 113) and the valley of the Oued Souss (page 88). This is also a biological crossroad between the steppic high plateaux to the east, the Jbel Siroua mountains to the north, the plain of the Oued Souss to the west, and the Anti-Atlas to the south. Taliouine is famous for the expensive, but very tasty spice 'safran' (the stamens of *Crocus sativus*), that can be bought at the 'Coopérative Souktana du Safran' (12 DH per gram in 2001).

Accommodation

Several small hotels and inns include the 'Safran' and 'Askaoun' and our two favourites: the 'Toubkal', with a swimming-pool (❸ ☎ 048.53.40.17), and the 'Auberge Souktana' (❸ ☎ 048.53.40.75); both also welcome camper vans and tents. The 4-star 'Ibn Toumert' (❶ ☎ 048.53.43.33, Fax 048.53.41.26) isn't currently good enough to justify its top rating and can't be recommended. Be aware that the region can be cold even late in spring, and warm clothes are sometimes necessary at night as the basic hotels have no heating (but they do have hot water).

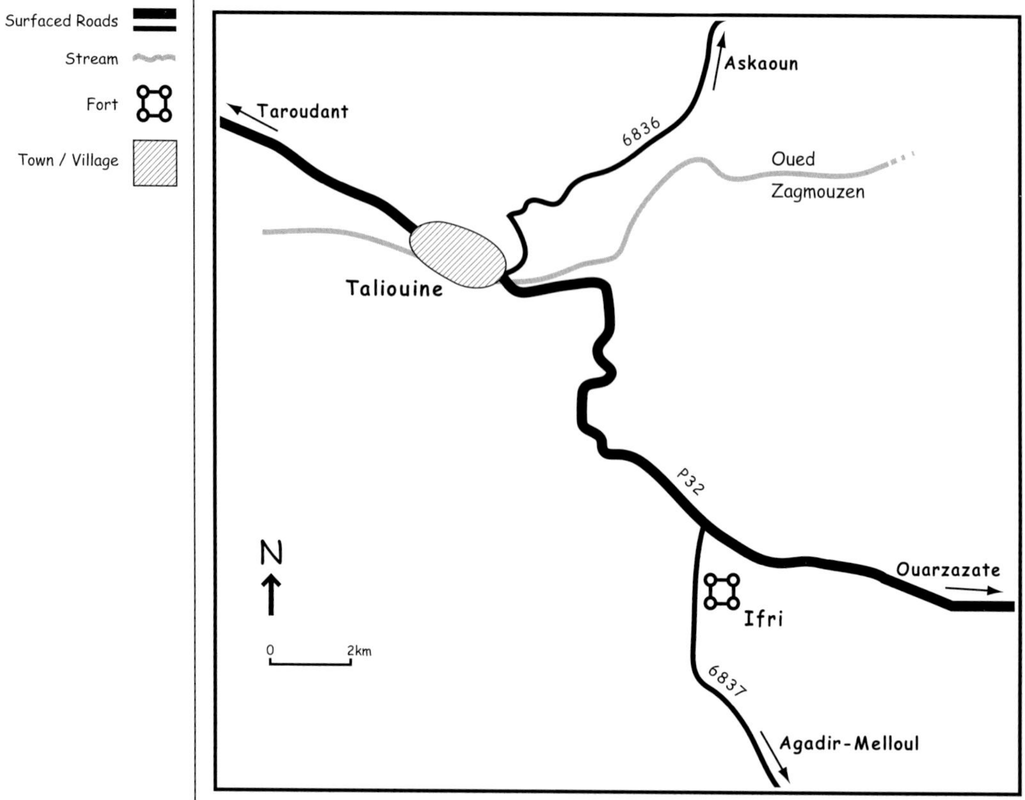

Strategy

The Oued Zagmouzen and the nearby vegetation and almond orchards are of special interest for birders, and can easily be reached on foot starting from the Auberge Souktana. They are at their best on April mornings, when many migrants are present as well as local species.

If time permits, it is a good idea to look for birds on the steppic high plateaux to the east of the town and visit the little village of Ifri (see page 95). People wanting to hike into the mountains start their ascent to the Jbel Siroua (3,304m) at Taliouine with the help of one of the few guides available, including Ahmed Jdid, the owner of the Auberge Souktana.

Birds

Migrants pass in good numbers in April. In the almond orchards you should see Turtle Dove, Black-eared Wheatear, Blackcap, Wood, Willow, Subalpine and Spectacled Warblers, and Woodchat Shrike. European Bee-eaters commonly overfly the area, but only a few stop here to breed. Grey Heron, Little Ringed Plover, Common, Green and Wood Sandpipers, Red-rumped Swallow, White (Moroccan race), Grey and Yellow Wagtails frequent the oued. The only Moroccan record of Greater Painted-snipe comes from Taliouine.

Breeding species in the orchards and bushes include Hoopoe, Common Bulbul, Cetti's and Sardinian Warblers, Rufous Bush Robin, Nightingale, Moussier's Redstart, Blackbird, Great Tit, and several finches. Red-rumped Swallow and House Bunting prefers the man-made constructions (they were both nesting at the Hotel Toubkal in 2001); Spotless Starling and Rock Sparrow breed in the old Kasbah near the Hotel Ibn Toumert, and Collared Dove is now common (first record in 1994).

The region is good for raptors, and likely species include Long-legged Buzzard, Booted Eagle, Peregrine Falcon, Kestrel, and Eagle, Tawny, Little and Scops Owls. Dark Chanting Goshawk, Black-shouldered Kite and Red-footed Falcon have also been recorded.

The steppes at the east of the town are home to larks (including Thick-billed), Trumpeter Finch and wheatears, with Black and Black-eared being the commonest. Other desert species have included Cream-coloured Courser, Tawny Pipit, and Scrub Warbler.

Others

Mauritanian Toad can be found in the Oued Zagmouzen.

Ifri

(9b – Central Anti-Atlas)

The small village of Ifri has a very impressive 'Agadir' (a collective, fortified store-house) – see also Amtoudi-Id Aïssa page 99 – and is really worth visiting if you are interested in the Berber culture. This is one of only a few that are still actively used in Morocco and has been there since time immemorial. It is used by the Berber community, and each family owns a small room where they store their valuables: the Agadir can be compared to a bank and the rooms to safes. The nearby cliff holds a large colony of Rock Sparrows and a pair of Peregrines.

Location

Ifri is some 16km from Taliouine (see map page 94); starting from Taliouine, follow the main P32 road to the east for 14km after the bridge over the Oued Zagmouzen. Then turn to the right on a track (which could well be asphalted soon) signposted 'Agadir-Melloul'; this is track 6837 of the Michelin map. Park your car 2.4km later near the first village on the left side of the track (Ifri). You will see the Agadir huddled against the cliff, just above the village.

Accommodation

There are no hotels or restaurants available at the site (fortunately!) and Taliouine is the nearest place to stay (see page 93).

Strategy

Because of the position of the Agadir which is sheltered against a roughly north-south cliff, the site is best visited in the afternoon when the sun lights it – nevertheless, avoid doing so in the hottest months when the temperature can become uncomfortably hot. Plan three or four hours for the visit, if possible in mid-spring.

When parking, tell any children who approach that you want to visit the Agadir and they will take you to the Cheikh (the Chief of the village) or his wife Aïcha who are in charge of the store-house; they will open it and act as guides during the visit. A powerful torch or flashlight is essential for seeing the internal rooms.

Remember the Tourist Code (see page 12) and please don't give anything to the children; if you want to help them, give something to the young schoolmistresses at the little school (such as paper, books, and pens). Do give a tip to Cheikh or Aïcha for acting as a guide though (DH20 was appropriate for a couple of hours visit in 2001).

If visiting in the spring, after seeing the Agadir, walk south along the cliff for the Rock Sparrows then go back to the car through the orchards and gardens looking for migrant passerines.

Birds

The most important feature of the site is an impressive Rock Sparrow colony located in a high cliff 100 metres or so to the south of the Agadir. There were more than 100 pairs breeding there in April 2001, and most of them were feeding young at the nest. A pair of Peregrine Falcons (*brookei*) and many Rock Doves were also present.

Other breeding species included Kestrel, Turtle Dove, Common Bulbul, Moussier's Redstart, Blackbird, Black Wheatear, Western Olivaceous Warbler, Great Tit, House Sparrow, Serin, and Goldfinch.

In spring the orchards and gardens attract migrants, and European Bee-eater, Crag Martin, Yellow Wagtail, Redstart, Northern Wheatear, Melodious, Subalpine and Willow Warblers, Pied Flycatchers, and Woodchat Shrike have all been recorded.

Other Wildlife

Some of the most remote rooms of the Agadir shelter large colonies of bats; Barbary Ground-squirrel is common everywhere in the cliff.

A trip in the Jbel Bani
(9 – Anti-Atlas)

The region to the south of the Anti-Atlas is basically comprised of immense tracts of stone desert, desolate cliffs, and mountain ranges (Jbel Bani and Jbel Ouarkziz), and there are only a few oases. The avifauna of these very typical desert habitats is still poorly known (the region has been open to tourists only since the late 1990s) and it is therefore well worth exploring.

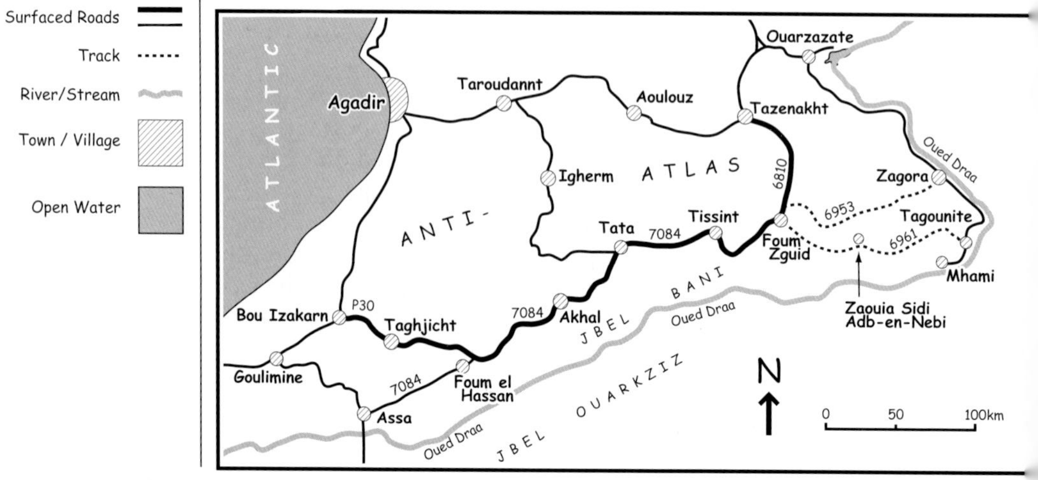

Accommodation

Little accommodation is available in this area and trips therefore have to be planned accordingly. In 2002, decent hotels were only available at Taghjicht and at Tata. There are no organised camp sites, but camping is permissible everywhere. At Taghjicht, the 'Auberge Touristique – Taregua' (❸ ☎ 048.78.87.80) seems best. At Tata, 'La Renaissance' (❸ / ❷ 9 Avenue des F.A.R. ☎ 048.80.20.42 and 20.25), and the 'Relais des Sables' are both good (❷ Avenue des F.A.R. ☎ 048.80.23.01, Fax 048.80.23.00). These hotels also serve good value meals, and wine is available with meals.

Strategy and Birds

The circuit described here, from Taghjicht to Tazenakht, can be done in two or three days though four days are needed to fully explore the region. An example of a 4-day trip is as follows, but many other alternatives exist. **Day 1:** visit the Taghjicht oases and Id-Aïssa, night at Taghjicht. **Day 2:** travel from Taghjicht to Tata, visiting oases en route, night at Tata. **Day 3:** visit oases near Tata, explore Tata to Tissint road and back to Tata for the night. **Day 4:** Tata to Tazenakht or Ouarzazate.

The best time to visit the area is probably the early spring, in March and April. From May onwards, temperature rises dramatically and birdwatching may become much more difficult and less productive then; the highest temperatures in Morocco have been recorded at Tata (up to 52°C). As for any other desert areas, it is best to start birding early in the morning, break in the middle of the day, and then start again late in the afternoon.

The region is famous for its countless rock carvings (2,000-500 BC); several sites are sign-posted along the road and a rough sketch map is available at the entrance of Hotel de la Renaissance at Tata. If these are of particular interest, it is best to arrange for a local 'guide' at Taghjicht, Akka or Tata (guides can be found at the hotels or cafés – be sure to agree a price before starting the trip).

As this has long been a sensitive region because of the war with the 'Front Polisario', there are still some military ('Gendarmerie') check points, such as near Icht and at Tissint (in 2002). Passports may be requested, but generally things go smoothly. Driving on the main roads and on the major tracks is safe and fairly easy but ask permission/ advice from the local authorities if intending to drive off road to the south as mines may be still present; the help of a local guide is advisable.

The major road crosses vast tracts of stone desert (called 'reg'); typical species include the easy-to-find Cream-coloured Courser, Hoopoe and Thekla Larks, Desert and Bar-tailed Desert Larks, Desert and White-crowned Black Wheatears, Southern Grey Shrike, and Trumpeter Finch. Black-bellied, Crowned and Spotted Sandgrouse are common but are often difficult to see away from drinking places; Fulvous Babbler is fairly common in bushes along the dry river beds and in some palm-groves. Tristram's Warbler is a fairly common winter visitor. Egyptian Nightjar, Desert Warbler, and Brown-necked Raven are rarer and may be more difficult to find. The numerous bare cliffs are the habitat of Bonelli's Eagle, Kestrel, Lanner, wild Rock Dove, and 'Pharaoh' Eagle Owl.

Stop in the oases to look for resident and migrant species, as they generally attract lots of birds during migration times. Breeding species include Laughing, Collared and Turtle Doves, Common Bulbul, Rufous Bush Robin, Blackbird, and Western Olivaceous Warbler. Serin and Goldfinch reach the southern limit of their breeding range here.

Collared Dove, Common Bulbul, House Sparrow, and House Bunting are common in most towns.

Taghjicht
(9a – Western Anti-Atlas)

Taghjicht is the westernmost oases in the Jbel Bani and its two main features are the palm-groves along the Oued Sayed (the same oued flows just to the south of Goulimine, see 'The West Sahara coast' page 124) and the Adrar-n-Saras mountain which overhangs the town.

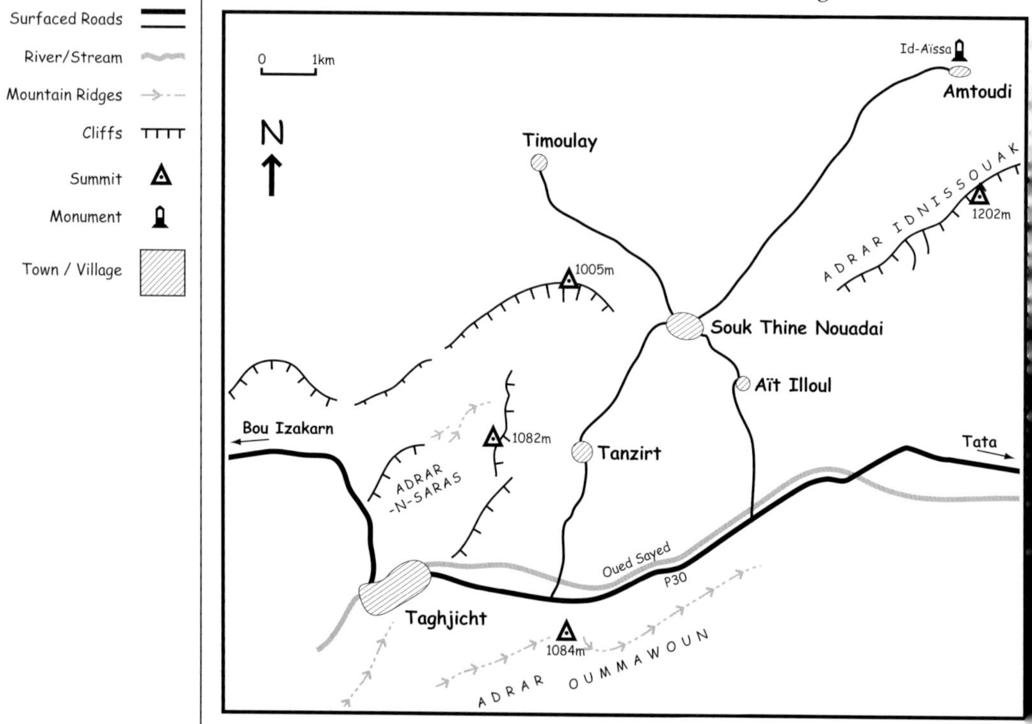

As elsewhere, the palm-groves attract lots of passerine migrants (for example, a large passage of Subalpine Warblers in early April 2000). Some waders stop along the banks (Grey Heron, Little Ringed Plover, Green Sandpiper). Kingfisher was present in April 1999 and 2000 and this could be one of the southern limits of this species' breeding range. Coming from the west, the first Laughing Doves will be encountered amongst the commoner Collared and Turtle Doves. On 10 May 1942, Heim de Balsac obtained a Namaqua Dove here; only two subsequent records are known in Morocco.

A pair of Bonelli's Eagles breed in the cliffs of the Adrar-n-Saras and they can sometimes be watched overflying or perched on the mountain, while having a cup of mint tea on the terrace of the Auberge. Climbing to the top of the Adrar (1,082m) is very rewarding for the superb scenery, the old 'castle' (possibly from the Almohavide period of the 10th century) and the birds which include Blue Rock Thrush, Desert Lark, Trumpeter Finch, and House Bunting. African Rock Martins breed in the high eastern cliffs of the Adrar.

Amtoudi – Id-Aïssa
(9a – Western Anti-Atlas)

Id Aïssa is a superb berber Agadir (a collective fortified store-house), and is one of the best preserved and most spectacular in Morocco. It was built some 800 years ago on a pinnacle of rock near the village of Amtoudi. Those interested in Berber culture and architecture shouldn't miss it. Some interesting bird species have been recorded here.

The site can be reached via a newly built small road (12 kilometres long) starting some 3km to the east of Taghjicht on the P30 (see 'Taghjicht and Amtoudi' map). Drive through the village of Tanzirt, then turn left in the village of Souk Tnine Nouadai to Amtoudi, which lies 7km further on. At Amtoudi, go through the village and park your car near the spring, under the welcome shade of large trees.

Climbing up to the Agadir takes half an hour on a winding path; alternatively, donkeys can be hired to ride. Hire one of the few local 'guides' who will open the Agadir and explain how it was operated (Ali is the official key-keeper, and accepted 40 DH for a 3-hour trip in 2000). By following the wonderful valley upstream for some 2.5km a series of 'gueltas' (natural pools) are reached where the local people bathe (but beware of the risk of bilharzia).

Drinks and light meals such as omelettes are available from the small café near the car park. On the return journey turn left at Souk Tnine Nouadai and go back to the main P30 through the village of Aït Illoul to complete the circuit.

In April 2000 Collared and Laughing Doves were not present, but Turtle Dove, Common Bulbul, Blackbird, and Western Olivaceous Warbler were singing in the tiny palm-grove of Amtoudi. Many flocks of European Bee-eaters were seen migrating north. House Bunting was common everywhere and from the top of the Agadir a Bonelli's Eagle, a pair of Common Ravens, Crag Martin, Blue Rock Thrush, and Black Wheatears were seen. This may be one of the few contact areas between Black and White-crowned Black Wheatears.

Between Taghjicht and Tata
(9a and 9b – Western and Central Anti-Atlas)

Several interesting spots are worth exploring between Taghjicht and Tata and a day should be allocated for driving this 208km section.

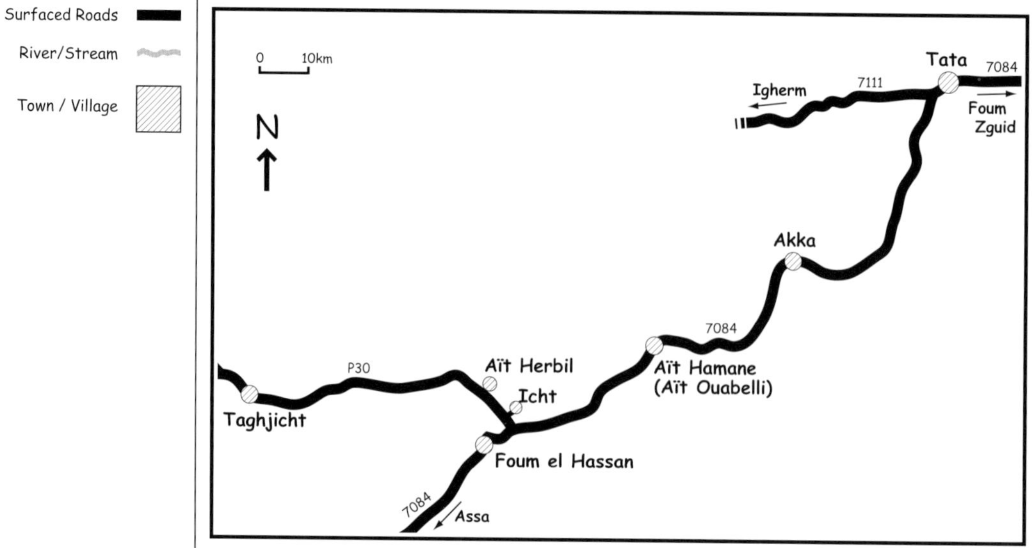

The first part of the road, between Taghjicht and Icht, crosses several dry riverbeds with scattered acacia and argana trees and bushes, which are good for Fulvous Babblers. Probable African Rock Martins and a road-killed Genet were seen at the Oued Tamanart near Aït Herbil in April 1999.

A few kilometres after Icht, a large new road (7084) on the right leads to Foum El Hassan (5km) then continues on to Assa (80km) through wonderful scattered acacia woods. In the opposite direction the palm-grove at Aït Hamane (also called Aït Ouabelli on the Michelin map) is good for Laughing Dove. Fulvous Babbler and Trumpeter Finch have been seen in the dry oued bed (Oued Aït Ouabelli) and Lanner has also been recorded here.

The small town of Akka (which has a market on Thursdays) is a convenient halt for a glass of mint tea en route. Though the town doesn't hold much of interest in itself, there is a nice palm-grove to the north with some pools where locals bathe (do not consider doing this because of the risk of bilharzia); the palm-grove is only accessible on foot.

28km before Tata, a small black, burned cliff on the left side of the road is a site for 'Pharaoh' Eagle Owl. In April 2000, a few odd pools were present at a small palm-grove on the Oued Tata, on the right side of the road 18km before the town. In addition to common migrants (including Little Ringed Plover and Green Sandpiper) a pair of Ruddy Shelducks were present and dozens of Spotted Sandgrouse came in to drink.

Tata
(9b – Central Anti-Atlas)

Tata occupies a strategic position in the Bani; it is the largest town and the only one besides Taghjicht that has hotels.

Just in front of the Hotel de la Renaissance, the Oued Tata is a good spot for spring migrants. In mid-April 1999 and 2000 birds seen included Grey Heron and Little Egret (an unusual sight in such desert landscape), Little Ringed Plover, Roller, large flocks of European Bee-eaters, House and Sand Martins, Yellow Wagtail, and Sedge Warbler. Other interesting species recorded here have included many waterbirds such as Squacco Heron, Collared Pratincole, Grey Plover, Little Stint, Dunlin, Snipe, Curlew, Green and Common Sandpipers, as well as Blue-cheeked Bee-eater, Pied Wagtail, Moustached, Western Bonelli's and Wood Warblers, Spotted and Pied Flycatchers, Golden Oriole, and Woodchat Shrike.

Some kilometres to the southwest of the town, along the 7084 road to Akka, there are some nice palm-groves. Migrants were plentiful here in mid-April 1999 and at the one named 'Tigzmerte' species such as Roller, Rufous Bush Robin, Nightingale, Subalpine and Willow Warblers, and Woodchat Shrike were seen.

Other possible species around Tata include Long-legged Buzzard, Crowned Sandgrouse (widespread) and Lichtenstein's Sandgrouse (localised to the south of the town near the Oued Draa; they favour the Acacia seyal trees and are more active at night), Egyptian Nightjar, 'Pharaoh' Eagle Owl, Fulvous Babbler (a small party were seen in a dry oued 5km to the east of Tata on the 7084 road) and Desert Warbler.

The landscape to the east of Tata along the 7084 road and then up the 6811 road to Foum Zguid, is basically the same as to the west; most of the palm-groves are worth a stop (for example, the authors had an African Rock Martin among migrating Barn and Red-rumped Swallows, and House and Sand Martins which were feeding and drinking at a basin in the palm-groves of Kasba-ej-Joua (19km before Tissint) in April 1999).

Tissint

(9b – Central Anti-Atlas)

Tissint is located 72km east of Tata and is one of the most spectacular spots of the circuit: water in the desert. Park on the left side of the road at the entrance to the village (close to the Gendarmerie check-point) and walk the short distance up to the edge of the cliff which overlooks the Oued Tissint and its waterfalls.

Rock Dove, Blue Rock Thrush, White-crowned Black Wheatear, and House Bunting breed in the cliffs that run along the Oued. Laughing Dove is common in the village.

The water attracts many other desert species and these have included Ruddy Shelduck, Bonelli's Eagle, Lanner, Barbary Falcon, Crowned, Spotted and Black-bellied Sandgrouse, Blue-cheeked Bee-eater, Temminck's Horned Lark, and Trumpeter Finch.

Many migrants are also attracted and watching Little Egret, Grey Heron, Little Ringed Plover, Wood and Common Sandpipers in this barren, desert landscape is always an unforgettable sight. Other migrants have included Short-toed Eagle, Osprey, European Bee-eater,

Common Swift, Barn Swallow, Crag and House Martins, Grey and Yellow Wagtails, Whinchat, Willow Warbler, and Spotted Flycatcher.

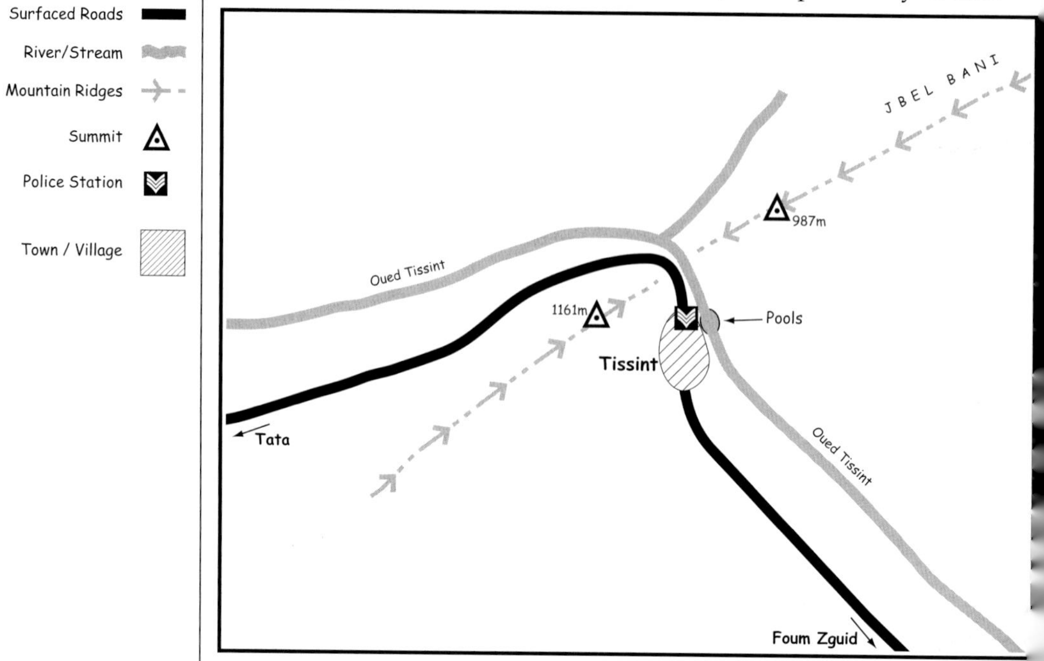

The course of the oued between Tissint and Mrhmina, some 15 km to the south-east of Tissint, is also well worth exploring, e.g. a flock of Lichtenstein's Sandgrouse was drinking at the oued 10km to the south-east of the town on 13 January 2002, 6.20 pm.

There is not much to see between Mrhmina and Foum Zguid, and this part of the trip can be done in one hour.

Foum Zguid and the road to Tazenakht

(9c – Eastern Anti-Atlas)

Foum Zguid is one of the 'major' towns of the region but up until 2002 it had no accommodation available except for in some cafés which provide very basic rooms. A walk in the large palm-grove should produce the usual list of resident (Collared and Laughing Doves, Common Bulbul, Rufous Bush Robin, Western Olivaceous Warbler, etc.) and migrant species.

There are three alternatives to end the trip. Two require a four-wheel drive vehicle: driving the 6953 track (120km) on the northern side of the Jbel Bani to reach Zagora in the Draa valley (see page 116) or driving the 6961 track (164km) on the southern side of the Bani to reach Tagounite, also in the Draa valley. The latter track is longer and more difficult but the birds might be rewarding as species such as Lichtenstein's Sandgrouse and Egyptian Nightjar have been recorded near the Zaouia Sidi-Abd-en-Nebi. In any case, seek advice from the local authorities before starting out on either of these two tracks.

If you have a regular vehicle, the only alternative is to drive north on the tarmac 6810 road to Tazenakht (100km) through wonderful landscapes; palm-groves go up to more than 1,000m, some 60km to the north of Foum Zguid, and then are replaced by rocky plateaux up to Tazenakht.

Tazenakht

(9c – Eastern Anti-Atlas)

Tazenakht is located at the strategic crossroads of the east-west main P32 (from Ouarzazate to the Souss valley) and the secondary S510 that leads to the Oasis of the Bani; it is thus a convenient stop. The town is famous for its carpets and rugs.

Accommodation

The city has several basic hotels among which the 'Zenaga' (❸ Main Road. ☎ 048.84.10.32) was the best choice in 2002. In addition, lots of open air restaurants serve cheap but delicious local food.

Strategy and Birds

Here again the landscape is basically comprised of rocky semi-desert ('reg') with typical birds including Lanner, Cream-coloured Courser, sandgrouse, larks, wheatears (including the scarce Mourning), Trumpeter Finch, etc. Tristram's Warbler has been recorded in winter.

The orchards and cultivations near the town on the road to Foum Zguid attract lots of migrants and are worth a visit, especially in the early morning. In April 1999 Hoopoe, several warblers including Melodious, Western Olivaceous and Orphean, Redstart, Woodchat Shrike, various finches, and large numbers of European Bee-eaters were present.

The Tagdilt Track

(9d – Sarhro)

This is an easily accessible steppe area to the south of the Central High Atlas where almost the full set of Moroccan desert birds can be found.

Location

The birding areas lie in the vast plain to the southeast of Boumalne du Dadès, between the southern slopes of the Central High Atlas and the Jbel Sarhro. Boumalne can be easily reached following the main P32 road either from the west (116km from Ouarzazate – see page 113) or from the east (190km from Er-Rachidia – see page 74); it is also the starting point for visits to the Gorges du Dadès (see page 71).

Accommodation

The closest accommodation is in Boumalne du Dadès and most birders stay at the 'Soleil bleu' (❸ ☎ 044.83.01.63, Fax 044.83.03.94), a good value hotel, which also has a bird-log describing the most recent sightings. The 'Soleil bleu' is in the eastern district of the city, to the north of the P32 and close to the 3* 'Madayek' (❶ ☎ 044.83.07.63 and 40.31, Fax 044.83.07.67 and 88.23.19). Other budget hotels include the 'Vallée des oiseaux' (❸ ☎ 044.83.07.64) and the 'Al-Manader' (❸ ☎ 044.83.01.72). The 'Chems' (❷ ☎ 044.83.00.41) and the 'Kasbah Tizzarouine' (❷ ☎ 044.83.06.90, Fax 044.83.02.56) are medium category and are good choices for those seeking greater comfort. Alternatively, one of the hotels and inns in the Gorges du Dadès could be used as a base (see page 71).

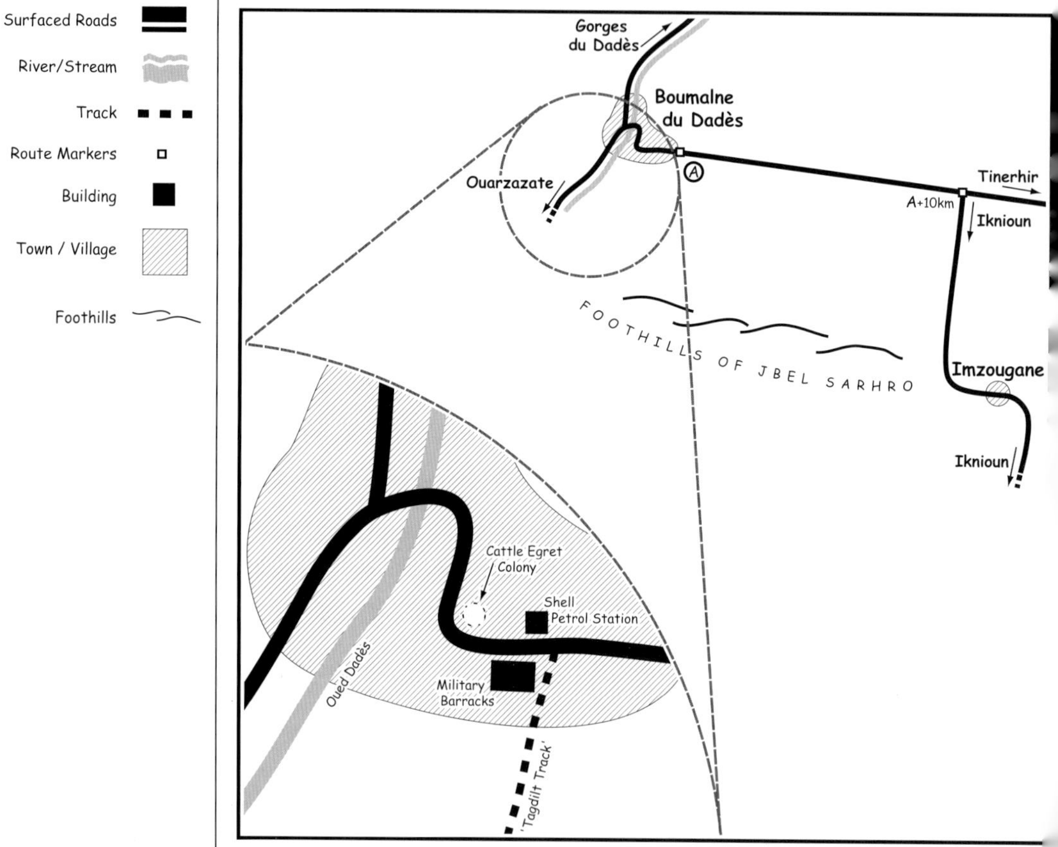

Strategy

The area is best visited in spring (but not before February-March if you wish to see Cream-coloured Courser) and in autumn when many migrants will be recorded in addition to the residents. Summer should be avoided because of the heat. Start birding early in the morning but take warm clothes as the area is at 1,500-1,600m and it can be surprisingly cold at times, even in late spring. It is worth staying on until late afternoon to enjoy the sunset in the desert.

The so-called Tagdilt track starts on the main P32 just to the east of the military barracks, in the eastern outskirts of the city close to a Shell petrol station. However, this is not a very pleasant area as it passes through rubbish dumps and there are many glass bottles, tins, and black plastic bags (a Moroccan plague) strewn across the ground, but it is the closest access point to town. An alternative is to drive some 10km east of Boumalne and take the road south towards Iknioun; many tracks run west from the Iknioun road and head towards the Jbel Sarhro, and they are usually as productive as the Tagdilt one, and are in a much more pleasant environment.

One of the best strategies, both for watching and photographing birds, is to drive slowly along the tracks making frequent stops to scan the steppe. The first few hours can be disappointing if you are not familiar with this kind of habitat and the secretive behaviour of some of its inhabitants and it is essential not to give up too quickly.

Birds

Raptors present include Long-legged Buzzard, Bonelli's Eagle, Lanner, Barbary Falcon and sometimes the African race (*minor*) of Peregrine. Many other species have been recorded including Sparrowhawk, Golden, Tawny, Short-toed and Booted Eagles, Griffon and Egyptian Vultures, Montagu's and Marsh Harriers, Kestrel, and Merlin.

One of the most sought-after birds in Morocco, Houbara Bustard, was fairly common here until recently but their numbers decreased dramatically due to frequent visits by hunting parties from the Arabian Gulf states and the latest known record dates from April 1999.

Cream-coloured Courser is a migrant breeder that leaves the region in August-September and returns in late winter (December to March). It has been suggested that these birds may breed once in the Sahel in autumn and again here in the spring but there is no evidence to support this as yet. This bird can be difficult to spot without experience but it is fairly common in spring.

Dotterel was only thought to winter in the steppe high plateaux of Eastern Morocco and in the plains of Central Morocco; it was only considered as an accidental visitor south of the High Atlas. However, there have been recent records here and this could indicate an unknown wintering area.

Sandgrouse often fly over the area in search of drinking sources and are best seen in the first few hours after sunrise. The commonest are Black-bellied and Pin-tailed. Crowned is fairly common in the Sarhro and has also been recorded here. Spotted is probably the rarest.

'Pharaoh' Eagle Owl has been recorded in the banks of the small oueds that cross the plain and is probably common in the region. Blue-cheeked Bee-eater doesn't breed here but is sometimes seen in spring, from March onwards.

Larks are among the commoner birds. Lesser Short-toed is a resident while Short-toed is a migrant breeder present from March through September-October. Both can be encountered in large numbers especially during migration in March-April. Thick-billed is much less common; it is considered a resident but it wanders according the climatic conditions; it breeds from mid-March through June. Thekla favours more rocky areas, in particular in the foothills of the Sarhro; the status of Crested is unclear. Temminck's Horned Lark is resident and not uncommon in the steppe; breeding starts in April. Desert and Bar-tailed Desert Larks are also resident; the latter favours the sandier areas. Hoopoe Lark is fairly common and is one of the most spectacular larks of the area with an amazing song and nuptial display. Other larks have included Skylark and Shore Lark (irregular winter visitors), Dupont's Lark (probably accidental), and Calandra Lark (accidental).

The Tagdilt area is also excellent for wheatears. Black-eared Wheatear and the European races of Northern Wheatear are only migrants, while the Moroccan race (*seebohmi*), which breeds in the Atlas, is an irregular winter visitor. Desert is a migrant breeder present from March through to October, though small numbers stay through the winter. Red-rumped is a resident; this is one of the commonest wheatears here and one of the most conspicuous, often first located via its 'boiling kettle' song. Mourning is also resident but this is one of the rarer species. Black occurs mainly in the rocky areas. Finally, Isabelline Wheatear is a rare spring passage migrant, mainly in March-April.

Common Raven has become rarer over the years probably due to widespread poisoning campaigns. Brown-necked Raven has been claimed here but there is no definite proof; it generally favours more desert areas further south.

Trumpeter Finch lives mainly in the foothills; in winter, Crimson-winged Finch and Red-billed Chough sometimes come down from the High Atlas to the Tagdilt steppe.

The few orchards and gardens attract lots of migrant passerines; there is a pleasant one at 10.5km from the P32 on the road south to Iknioun (some 800m before the first village, Imzougane). During a late April visit Redstart, Whinchat, Northern Wheatear, Blackcap, Pied Flycatcher, and Woodchat Shrike were recorded, while Common Bulbul, Western Olivaceous Warbler, Blackbird, and Goldfinch were likely breeders.

Collared Dove was first recorded at Boumalne du Dadès in 1997; this is now a common species. A small colony of Cattle Egrets breeds on a pine intown, on the northern side of the road just before driving down to the Oued Dadès; this is one of the southernmost colonies.

Other Wildlife

North African Green Frog and Mauritanian Toad are common in the Oued Dadès. Changeable Agama favours the rocky areas. The steppe is punctuated with rodent burrows and several Lesser Egyptian Jerboas have been found dead on the road; larger mammals have included Common Red Fox and Jackal.

The Tafilalt: Erfoud, the Erg Chebbi and Dayet Merzouga

(10b – Tafilalt)

The Tafilalt is the region in the southeast of Morocco closest to the Sahara, centred on the towns of Erfoud and Rissani. Once famous for its trade in gold, ivory and ebony, it nowadays offers birdwatchers an opportunity to see a wide range of Saharan desert species. At the southwest end of the sand dunes of the Erg Chebbi, Dayet (lake)

Merzouga (also called Dayet Srij) attracts thousands of birds during wet seasons and then forms an unforgettable sight.

Location

The region lies very close to the Algerian border in the southeast of Morocco, where the main roads from the north and west come to an end.

There are three routes leading to Erfoud: the main P21 that comes from the north, via Er-Rachidia and follows the Oued Ziz, the 3451 that starts at Tinejdad on the P32, 55km east of Tinerhir, and the 3454 that starts in the Draa valley. These roads pass through attractive landscapes and, if travelling by car, it is worth arriving by one route and leaving by another.

Erfoud can also be reached by bus from Er-Rachidia. A Land Rover and driver can be hired for a half- or one-day trip out into the desert. Land Rovers can be arranged through hotel reception desks.

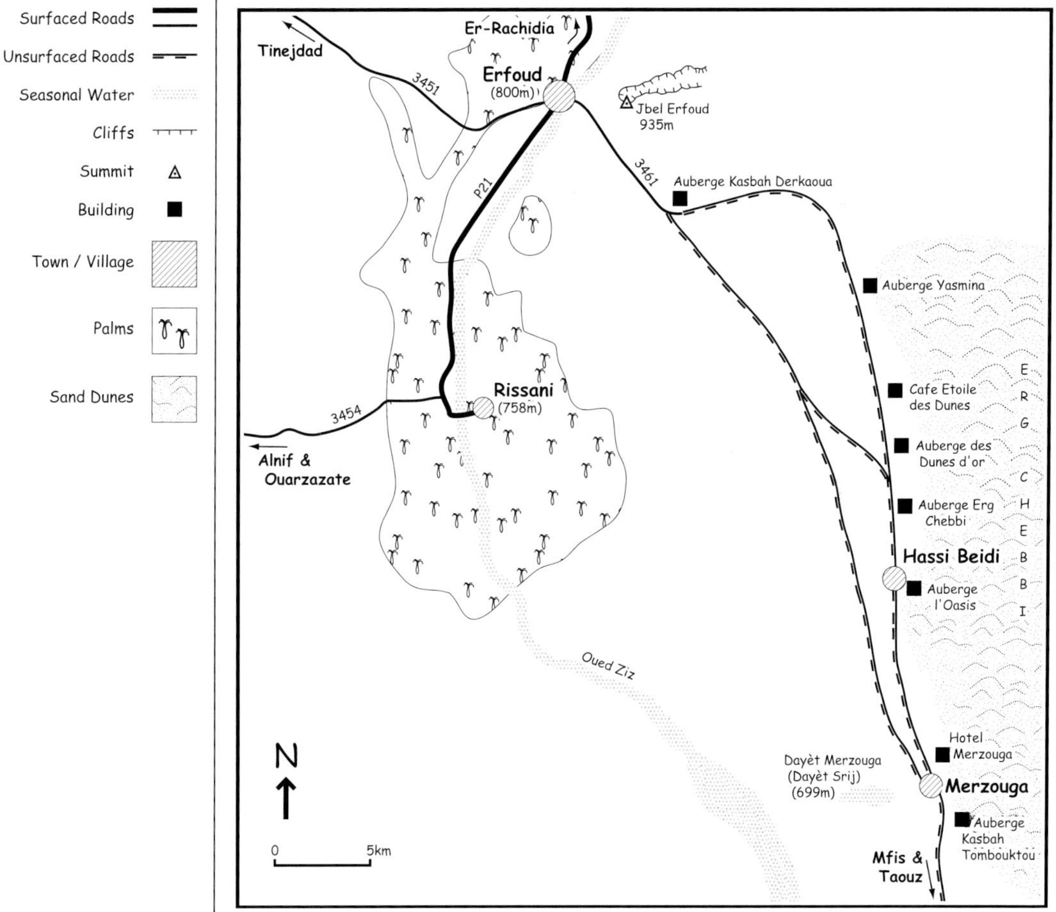

Accommodation

There are now many hotels in Erfoud and near the Erg Chebbi. The medium and high category ones are at Erfoud; they are often fully booked in spring and it is wise to reserve a room in advance. A lot of basic hotels have been built close to the Erg Chebbi (the map only shows some of them), and any one of these is a good base for those willing to accept less comfort, but with the advantage of staying right in the desert. In Erfoud the 'Ziz' (❷ 3 Avenue Mohamed V. ☎ 055.57.61.54, Fax 055.57.68.11) and the 'Kasbah Tizimi' (❶ / ❷ Route de Tinerhir. ☎ 055.57.61.79 and 73.74, Fax 055.57.73.75) are recommended in the medium category. Close to the Erg Chebbi at Merzouga the 'Auberge des Dunes d'or' (❸ ☎ 061.35.06.65) and the 'Auberge Kasbah Tombouktou' (❸ ☎ 055.57.67.93, Fax 055.57.77.65) can be recommended, but many others of approximately the same standard are available. If money is no problem try the 'Auberge chez Michel' (also called 'Kasbah Derkaoua' ❶ ☎ and Fax 055.57.71.40) between Erfoud and Merzouga; if staying overnight here, look in the gardens (which attract lots of migrants in spring and autumn) and also in the nearby dry wadi (Spotted Sandgrouse, Egyptian Nightjar, Tristram's Warbler, Fulvous Babbler, and Desert Sparrow have been recorded). There is a very modest camp site at Erfoud, and another between Erfoud and Er-Rachidia, at Source bleue de Meski. Most of the inns near the Erg Chebbi also welcome campers.

Strategy

A full day is necessary to explore the Tafilalt thoroughly. March to May is probably the best period to visit this region. October to Februar can also be interesting, but the summer visitors are absent. Avoid the period from June to September when it is very hot and uncomfortable for birdwatching. In spring, especially in May, it is best to start birdwatching at dawn as by mid-morning it can be quite hot, the birds are less mobile, and the heat haze becomes a problem. On hot days it i a good idea to rest in the middle of the day and to start birdwatching again in the late afternoon.

The three main habitats in the region are palm-groves, stone deserts, and sand dunes. Each of them contains their own distinct avifauna. Palm-groves are found all along the Oued Ziz; the Erg Chebbi is the only major sand dune system in Morocco and is situated to the east of Merzouga; everywhere else is stone desert.

Numerous tracks cross the desert but there is little risk of getting lo Nevertheless, before setting out, check petrol levels and the condition of the spare wheel and travel with another car if possible (it is no fun breaking down alone in a remote area). When there is a sandstorm do not venture out as it is then very easy to get lost.

One of the best tracks to explore is the 3461 which leads from Erfou to Merzouga and the Dayet Srij; it is tarmac for some 15 kilometres an then splits into several tracks some 35km from Merzouga. Follow the line of telephone poles that serve the village of Merzouga and you wi soon see the north-south line of the sand dunes. With a 2-wheel drive vehicle, it is best to avoid the cross-tracks that appear on most maps between Merzouga and Rissani. Don't drive too far south as the area close to the Algerian border is rather sensitive. It is best to stay within the limits of the map (see page 107).

Even those with their own transport should consider renting a Lan Rover and driver in Erfoud for their first trip into the desert. It is then wise to arrange for an experienced and reliable driver through one of the main hotels; don't trust the numerous 'guides' on the streets in Erfoud who promise to show visitors bustards.

Birds

The palm-groves and adjacent areas along the Oued Ziz hold birds such as Barbary Partridge, Moorhen, Turtle Dove, Scops and Tawny Owls, Hoopoe, the Moroccan race of White Wagtail, Common Bulbul, Rufous Bush Robin, Cetti's and Western Olivaceous Warblers, and Spotted Flycatcher during the breeding season. One of the most sought-after birds, Blue-cheeked Bee-eater is fairly common from April through September. European Bee-eater also breeds and passes through in large numbers. In migration periods, numerous birds stop to feed, and these include species such as Spotted, Little and Baillon's Crakes, and Roller.

The dunes of the Erg Chebbi are relatively poor for birds but the landscape is unique in Morocco and worth experiencing. Only six species breed here; Bar-tailed Desert, Short-toed, Crested and Hoopoe Larks, Desert Warbler, and Desert Sparrow. Desert Sparrow is best looked for around one of the various inns close to the dunes, and especially at the northernmost ones, e.g. at the 'Auberge Yasmina' where it breeds. The tamarisks that fringe the dunes at Merzouga are a very good place to observe wintering Tristram's Warblers from mid-November to late April (mainly from December to February).

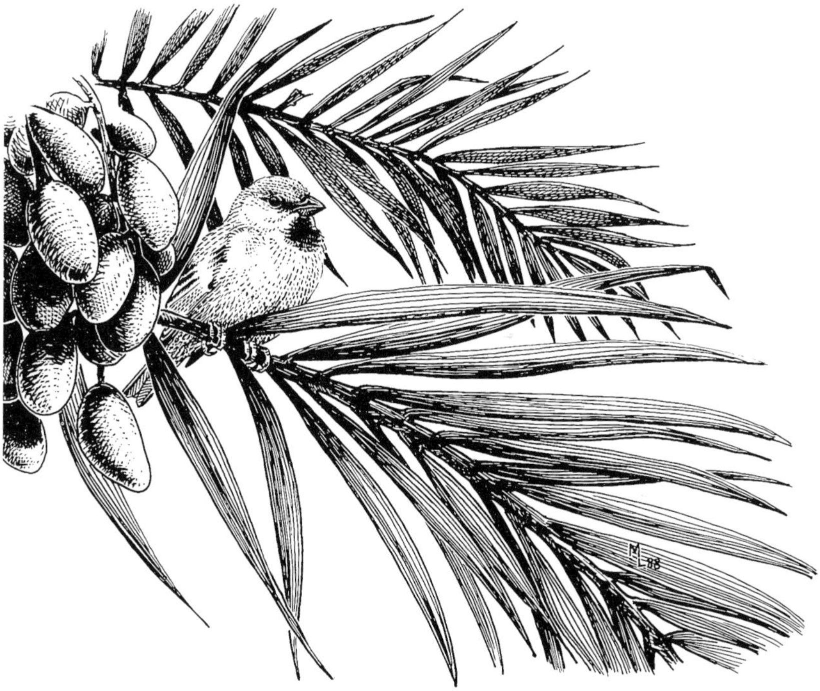

Desert Sparrow

Although some birds are very conspicuous, many of the species that inhabit the vast tracts of stone desert can be difficult to find. Regular breeding birds include Stone-curlew, Spotted Sandgrouse, Bar-tailed Desert, Desert and Hoopoe Larks, Desert, Mourning and White-crowned Black Wheatears, and Trumpeter Finch. Houbara Bustard has suffered a dramatic decline because of hunting parties of falconers from the Gulf States (for example, 117 birds were killed in January 1984 and 120 in December 1988), and has now almost disappeared. There have been a few reports of Arabian Bustards in the desert south of Merzouga several times since the 1980s, from September through to April, and these are most probably African *stieberi* birds. Cream-coloured Courser, Pin-tailed Sandgrouse, Thick-billed, Short-toed and Temminck's Horned Larks, and Spectacled Warbler only breed in certain years, depending on climatic conditions.

The small dry wadis covered with vegetation are home to Desert Warbler (the area some 5-6km from Erfoud towards Merzouga has proved to be excellent for this species), Scrub Warbler, and Egyptian Nightjar. Fulvous Babbler is a fairly common resident in the dry wadis scattered with jujube and acacia trees.

Long-legged Buzzard, Bonelli's Eagle, Kestrel, Barbary Falcon, Peregrine (African race *minor*) and Lanner breed in the region. Short-toed Eagle are regularly observed in spring, but breeding has never been proved. Honey-buzzard, Black Kite, Montagu's Harrier, Booted Eagle, Osprey, and Hobby are regular passage migrants, and Marsh Harrier and Sparrowhawk regularly winter.

Species which can be found in gardens, palm-groves, cliffs or stone desert include Barn, 'Pharaoh' Eagle and Little Owls, Blue-cheeked and European Bee-eaters, Crag Martin, Fulvous Babbler, Southern Grey Shrike, and Common Raven. Brown-necked Raven has become more common since the late 1990s (perhaps due to the effect of the drought) and flocks of a few dozen birds have been recorded near Merzouga.

If there is water in the Dayet Merzouga, try to arrive there at sunrise. White Stork, Greater Flamingo, Glossy Ibis, Ruddy Shelduck (sometimes several hundred birds), Wigeon, Gadwall, Common Teal, Mallard, Pintail, Marbled Duck (up to 3,000 in May 1973), Avocet, Collared Pratincole, Ringed and Little Ringed Plovers, Red Knot, Sanderling, Dunlin, Little and Temminck's Stints, Curlew, Green and Common Sandpipers, Ruff, Bar-tailed and Black-tailed Godwits, Spotted Redshank, Redshank, Greenshank, and Whiskered and Black Terns can all be seen and the presence of so many waterbirds in the open desert is an unforgettable sight. Black-winged Stilt, Kentish Plover and Gull-billed Tern have been noted breeding here during particular wet years.

Several rare species have been noted at the Dayet, including Great Egret, a 'dark' egret (either Western Reef Egret or a dark morph of Little Egret), Lesser Flamingo, American Wigeon, Lesser Spotted Eagle, Kittlitz's Plover, Wilson's Phalarope, Richard's Pipit, Isabelline Wheatear, and Collared Flycatcher.

Other Wildlife

Several reptiles will probably be seen and the most conspicuous are Bibron's, Changeable and Spiny-tailed Agamas.

Tamdakht

(10c – Dadès-Draa)

The kasbah (fortified village/castle) of Tamdakht (pronounced 'Tamdart') is a superb example of berber architecture. Its gardens cre a tiny spot of greenness in the surrounding black and red stone desert and attract lots of passerines during migration periods.

Location

Tamdakht is located some 35km from Ouarzazate in the southern foothills of the High Atlas, 5km to the north of the famous village of Aït Benhaddou which is included in the UNESCO World Heritage list. Don't be put off by the collapsed bridge over the Oued el Maleh 1km before Tamdakht (which will be rebuilt soon); there is generally only a little water in the oued and driving through it is usually no problem.

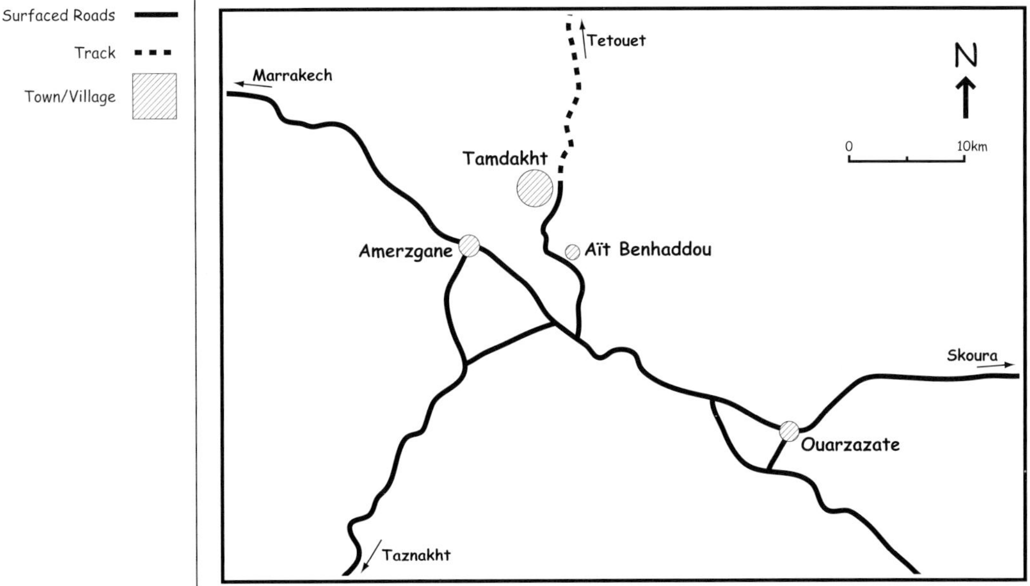

Accommodation

There are two options in Tamdakht itself. The 'Auberge des Cigognes' (❸ ☎ and Fax 044.89.03.71) is very basic, so the best choice is the 'Kasbah Ellouze' ('The House of the Almonds' ❷ ☎ and Fax 044.89.04.59, GSM 067.96.54.83. www.kasbahellouze.com). The latter has been built recently in traditional style, close to the old kasbah and overlooks the gardens, so is the ideal place to stay. Other hotels are available at Aït Benhaddou and include the 'Etoile filante d'Or' (❸ ☎044.89.03.22, Fax 044.88.61.13) and 'La Baraka' (❸ ☎ 044.89.03.05, Fax 044.88.62.73), both offering good value for money.

Strategy

Tamdakht means something like 'Administrative Centre' in berber. Nowadays, it is no longer an administrative centre and is only a small village of a hundred houses set around a wonderful berber castle (with only one inhabitant in it in 2002). It is well worthwhile seeking permission to view the castle.

From an ornithological point of view, two areas are of special interest here; the Oued el Maleh ('the salted river') that runs close to the village and is followed by many species during their migration, and the gardens that act as a magnet for migrating passerines. The perfect time for birdwatching here is in the early morning, especially in springtime when it makes a relaxing contrast after spending time in the desert. A good plan is to reach Tamdakht in the late evening, sleep at 'Kasbah

Ellouze' and spend the next morning in the gardens and along the oue (allow a couple of hours).

For those staying elsewhere and simply visiting Tamdakht during th day, cars may be parked close to the 'football pitch' a hundred metres before the village. From there walk along the path that runs close to th cemetery and leads to the oued.

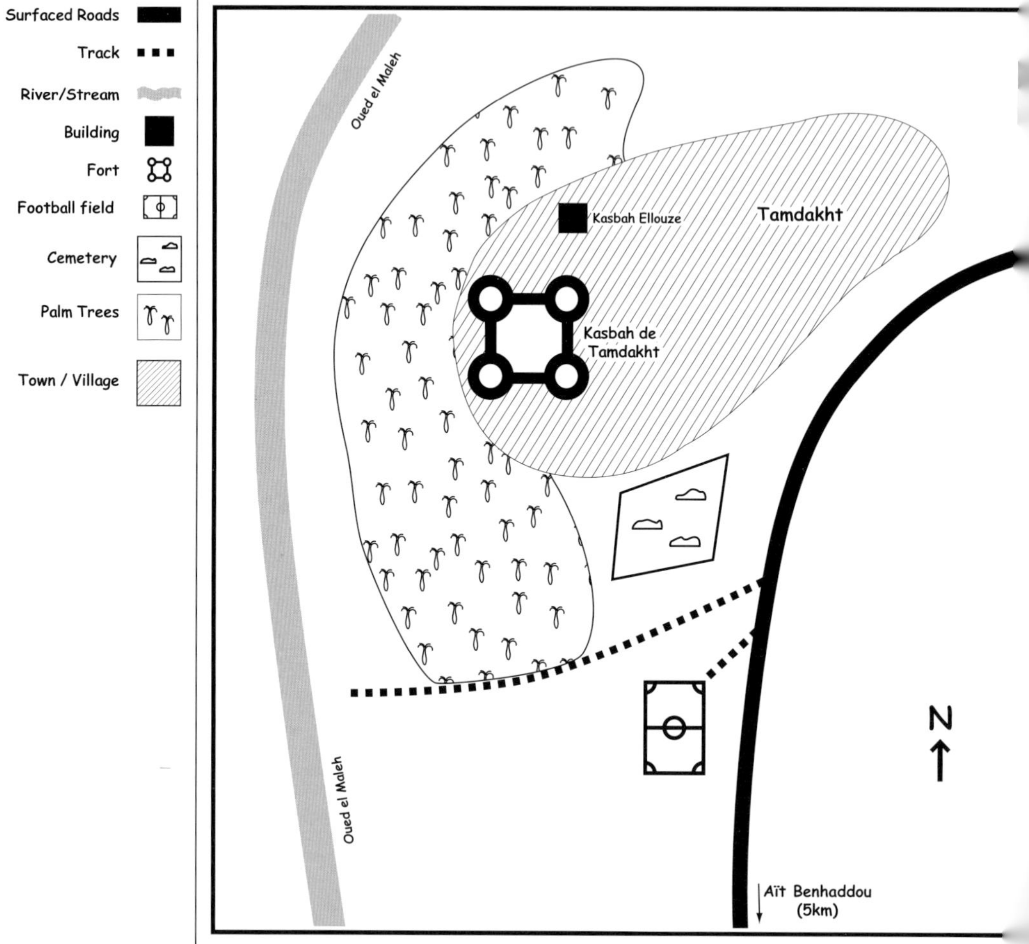

Birds Several pairs of White Storks breed on the kasbah walls; the famou Bald Ibis also bred on the crenels until 1976. House Bunting is comm in the village and Blue Rock Thrush sometimes perches on the kasbal

Regular breeders in the gardens include Turtle Dove, Hoopoe, Common Bulbul, Rufous Bush Robin, Nightingale, Blackbird, Cetti's, Sardinian and Western Olivaceous Warblers, Great Tit, Chaffinch, Serin, and Goldfinch. Laughing Dove has been discovered recently; this being its highest known altitude in Morocco (Tamdakht is at 1,400m). Collared Dove arrived in 2001. Little Ringed Plover, Grey and Moroccan White Wagtails breed along the oued.

In spring and autumn, passerine migrants can be plentiful. Late March and mid-April visits have produced Tawny and Tree Pipits, Rufous Bush Robin, Redstart, Northern Wheatear (migrant and Moroccan races), Blackcap, Subalpine and Western Bonelli's Warblers

Chiffchaff and Iberian Chiffchaff, Woodchat Shrike, and Golden Oriole in the gardens. In or near the oued Green Sandpiper, Pallid Swift, European Bee-eater, large flocks of Hirundines (including many Crag Martins and Red-rumped Swallows), Meadow Pipit, and Yellow Wagtail (including 2 *cinereocapilla*) have been seen.

Long-legged Buzzard, Black Wheatear, and Blue Rock Thrush are among the typical inhabitants of the cliffs. Desert and Thekla Larks, Mourning and White-crowned Black Wheatears, and Trumpeter Finch are typical of the stone desert.

Ouarzazate and the Mansour Eddahbi Dam

(10c – Dadès-Draa)

To the east of Ouarzazate lies the Mansour Eddahbi Dam which collects water from various rivers, in particular the Dadès, to form a large reservoir of variable extent.

The site is interesting not only because the dam forms the only large and permanent body of water in the sub-desert area south of the central High Atlas, but also because several desert birds can be observed nearby. The area around the lake, which appears dry and stony, with only a few bushes, shelters a surprisingly rich bird community.

Location

Ouarzazate lies on the crossroads where the road along the southern side of the High Atlas, from Agadir to Figuig (P32) crosses the P31 which runs from Marrakech to the Draa valley. It has become a popular tourist destination, with an international airport.

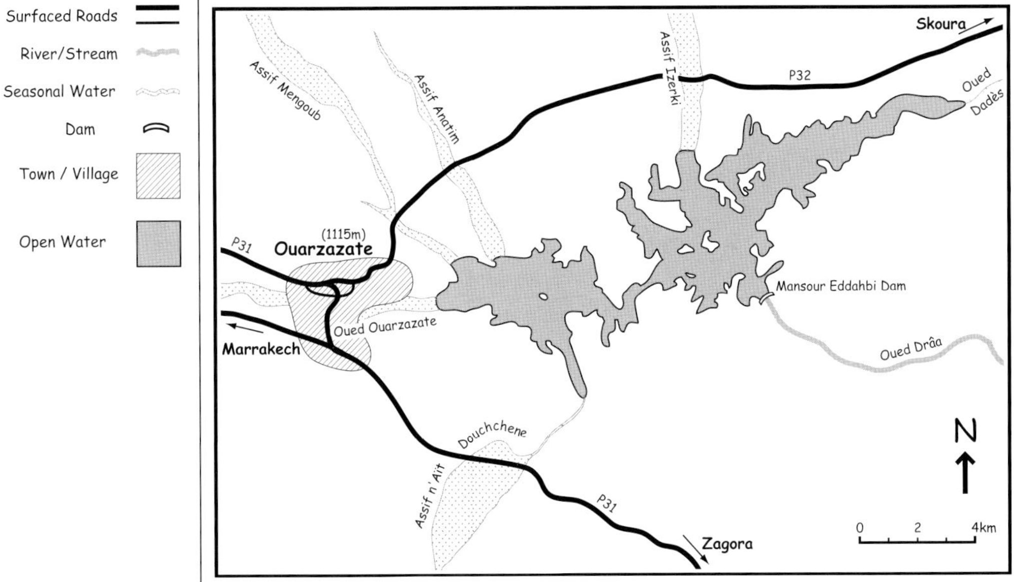

Accommodation

Ouarzazate has a wide variety of hotels. Some of the best value-for-money have been recently built in the new 'Tabounte' district, south of the Oued Ouarzazate on the road to Zagora. In 2002 these included the 'Sagho' (❸ ☎ 044.85.41.35, Fax 044.85.47.09), 'La Vallée' (❸ ☎ 044.85.40.34, Fax 044.85.40.43) and 'Mabrouka' (❷ ☎ 044.85.48.61, Fax 044.85.44.43), all three of which have swimming pools, and the latter air conditioning. A fallback choice in the same price category but downtown is 'La Gazelle' (❸ Avenue Mohamed V. ☎ 044.88.21.51). Several 4 to 5-star hotels are also available. Be sure to book in advance especially during the Easter holiday period. Restaurants are fairly numerous, and those in the above hotels are good choices.

Strategy

Because of its geographical position, Ouarzazate is a strategic overnight stop when travelling through southern Morocco. March and April are the best months for migrants and May is best for breeding birds. Avoid travelling in this region between June and September, because of the heat.

Try to arrive here in the afternoon as the evening can be interesting, when migrants stop off before crossing the Atlas the next morning. The following morning drive east on the P32 and explore the various tracks down to the reservoir. The best places for birds vary depending on the level of water. Closer to the reservoir beware of soft mud which can sometimes be invisible under dry surface crusts, especially when water levels are falling.

The north shore of the reservoir can be easily reached from the P32 along several roads and tracks, which are shown on the following map (the distances have been recorded using the odometer of a rented car and may be slightly out). Access to the southern side, from the P31, is more difficult and access to the dam itself is forbidden for security reasons. Another access point is from the east end of town, near the campsite (signposted 'Aït Kdif').

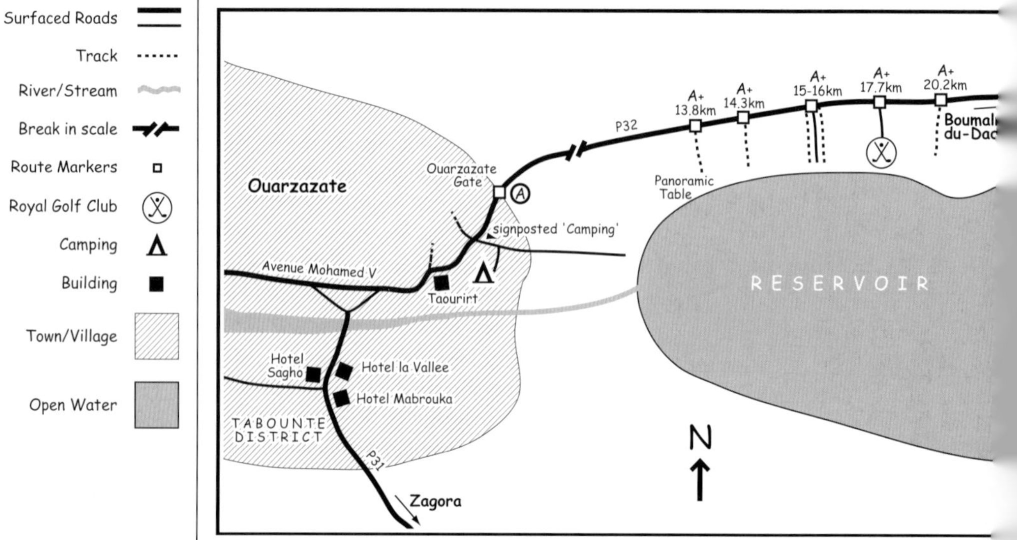

Be sure to look for passerines in the desert areas around the reserv

Birds

Large waterbirds here have included Little Egret, Squacco Heron, Black Stork (up to 25 in September 1983), Glossy Ibis, Spoonbill, and

Greater Flamingo. White Stork is common. A colony of Cattle Egrets breeds on the trees of the military barracks downtown. Three of the four colonies of Bald Ibis of the region vanished in the mid-1970s, the fourth one in 1984. The lake is the only breeding site for Grey Heron in Morocco, and Great Crested Grebe also breeds.

This is a good roosting and feeding place for wintering and migrating ducks, including Wigeon, Gadwall, Common Teal, Pintail, Garganey, and Northern Shoveler. Marbled Duck can be fairly common on migration, with up to 255 in April 1997. Shelduck, Pochard and Tufted Duck are rarer (but 111 Shelducks were present in November 1998). Greylag Goose is only an accidental visitor during harsh winters. Mallard and Ruddy Shelduck are the only two breeding species, and the latter can be common in spring (up to 500 in 1997).

Waders can be common on migration and during the winter. The species most commonly reported are Black-winged Stilt, Avocet, Collared Pratincole, Ringed and Little Ringed Plovers, Little and Temminck's Stints, Curlew Sandpiper, Dunlin, Green, Wood and Common Sandpipers, Ruff, Common Snipe, Black-tailed Godwit, Redshank, Spotted Redshank, and Greenshank. Others are much rarer, with only a few records of Grey Plover, Northern Lapwing, Red Knot, Sanderling, Bar-tailed Godwit, Whimbrel, Curlew, Marsh Sandpiper, Jack Snipe, and Grey Phalarope. Kentish Plover is the only breeding wader.

Only three gulls and terns are of regular occurrence: Black-headed Gull in winter, Gull-billed Tern in spring, and Black Tern during spring and autumn. Eleven other species have been recorded: Mediterranean, Little, Slender-billed, Audouin's, Lesser Black-backed, and Yellow-legged Gulls, and Common, Arctic, Little, Whiskered, and White-winged Black Terns.

Resident raptors include Long-legged Buzzard, Bonelli's Eagle, Kestrel, Lanner, Peregrine (African race *minor*), and Barbary Falcon. Sparrowhawk, Marsh and Hen Harriers, and Merlin are winter visitors. Honey-buzzard, Booted and Short-toed Eagles, Black Kite, Montagu's Harrier, and Lesser Kestrel are passage migrants. Osprey has been commonly recorded from autumn through to spring. Golden Eagle sometimes hunts over the area.

European Bee-eaters are among the commonest, more conspicuous migrant species in spring. Look carefully in the vegetation that fringes the small wadis for a wide range of migrants, including Little Bittern, Quail, Spotted and Baillon's Crakes, Red-throated Pipit, Orphean Warbler, and Woodchat Shrike. The local race of the White Wagtail (*subpersonata*) breeds. Tawny Pipit, Rock and Blue Rock Thrushes, Seebohm's and Black-eared Wheatears are best searched for in bare areas.

Among desert-loving species are Stone-curlew, 'Pharaoh' Eagle Owl, Black-bellied, Crowned, Spotted and Pin-tailed Sandgrouses, Desert and Bar-tailed Desert Larks, Desert and White-crowned Black Wheatear, and Trumpeter Finch. Thick-billed and Hoopoe Larks, and Mourning Wheatear can sometimes be seen here. This is also a good place for Blue-cheeked Bee-eaters (from early April) and Fulvous Babblers.

Laughing Dove, Rufous Bush Robin, Western Olivaceous Warbler, and Blue and Great Tits (local races) breed in orchards and palm-groves. Laughing Dove is still rare at Ouarzazate but is commoner southwards, in the Draa valley.

Rarities have included White Pelican, Great Egret, Dark Chanting Goshawk, Red-footed Falcon, Saker, Spotted Crake, Broad-billed Sandpiper, and Isabelline Wheatear.

The Draa valley
(10c – Dadès-Draa)

The Oued Draa starts at the Mansour Eddahbi Dam near Ouarzazate (see page 113) and then irrigates a splendid valley before vanishing under the Draa hammada near Mhamid, and re-appearing several hundred kilometres later to flow into the Atlantic (see the West Sahara Coast, page 121). A series of oases, palm-groves and 'ksours' (fortified berber villages) stretch along most of its course down to Mhamid and the valley is worth exploring both for the landscape and the birds.

The Barrage Mansour Eddahbi irrigation scheme.

The Mansour Eddahbi Dam was built between 1969 and 1972 to create a large reservoir (maximum capacity 536 million cubic metres) primarily to help avoid the terrible droughts which were a regular feature in the Draa valley as far south as Mhamid, and also for electricity production (10 MW). During normal years when there is enough water in the reservoir – from October through May approximately – the Authorities release water every two or three months, and these releases are spread over some 20 days (10 days or so are necessary for the water to reach Zagora). The Oued Draa then becomes a real river and a well-designed irrigation system allows all the palm-groves to be watered, down as far as Mhamid. These regular floods are of course greatly welcomed by the inhabitants and also by birds.

In early spring 2002, after some three years of drought, the reservoir was only filled to 12% of its capacity and there had been only two releases in the previous 12 months, the first in April 2001 and the second in March 2002. A mixed feeling of joy and frustration was perceptible in the local population. This reinforced the need for visitors to try to save water in the Moroccan pre-desert regions.

Location In the first part of its course, the Draa flows from northwest to southeast and then suddenly turns to flow west-southwest (the 'Draa elbow') near Mhamid. The main P31 road starts at Ouarzazate, crosses the dry and rocky Jbel Sarhro – the eastern part of the Anti-Atlas – down to Agdz (70km), then runs along the Draa down past Zagora (165km) and finally goes deep into the desert to the village of Mhamid (265km). This little town lies at the end of the tarmac road and it is th necessary to drive back to Ouarzazate on the same road. Alternatively, the 6956 road from Tansikht can be taken to the Tafilalt region (see page 106) or for those with a four-wheel drive one of the tracks 6961 or 6953 can be followed to Foum Zguid (see page 102).

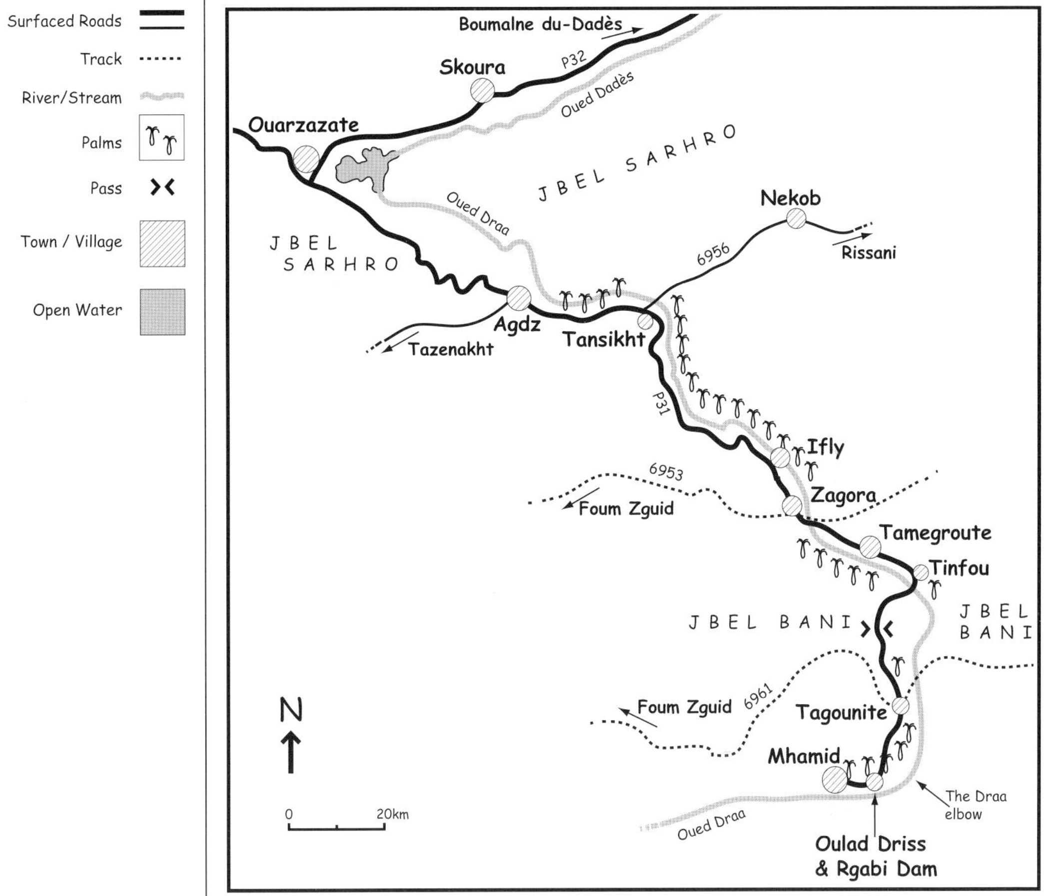

Accommodation

The Draa valley has become an increasingly popular tourist destination and many hotels and restaurants can be found in most towns and even in more remote areas. Before booking in to a hotel be sure to check for the availability of heating in winter and of air conditioning from May through to September. Perhaps here more than anywhere else be aware of conserving water in your hotel room as the region regularly suffers from severe droughts.

The following lists only includes a selection of the hotels available.

At Agdz, the 'Hotel des Palmiers' (❸ ☎ 044.84.31.27 and 31.87) is very basic, but the 'Kissane' is much more comfortable (❷ ☎ 044.84.30.44, Fax 044.84.32.58). There is a good campsite which also provides nice rooms at the 'Hotel Camping Jardin Tamnougalt' (❸ ☎ and Fax 044.84.36.14) 2km beyond the city, close to the Oued Draa under palm trees.

Zagora is a larger and a more pleasant city and is probably the best place to stay. Budget hotels include 'Hotel des Amis' (❸ ☎

044.84.79.24) and 'Vallée du Draa' (❸ ☎ 044.84.72.10), both on Avenu Mohamed V. Good choices in the middle category include the 'Palmeraie' (❷ ☎ 044.84.70.08, Fax 044.84.78.78) and the 'Reda' (❷ ☎ 044.84.70.79, Fax 044.84.70.12) before crossing the Oued Draa, and 'Kasbah Asmaa' (❷ ☎ 044.84.72.41 and 75.99, Fax 044.84.75.27), 'La Fibule du Draa' (❷ ☎ 044.84.73.18, Fax 044.84.72.71), 'Sirocco' (❷ ☎ 044.84.61.25, Fax 044.84.61.26) and 'La Perle du Draa' (❷ ☎ 044.84.62.10, Fax 044.84.62.09) after the bridge over the Draa. The 'Ksa Tinzouline' (❶ ☎ 044.84.72.52, Fax 044.84.70.42) is among the finest and more expensive ones. 'Chez Ali' (❸ Avenue Atlas, Zaouiat el Baraka ☎ 044.84.62.58), is a delightful little downtown inn with a superb tiny garden that attracts migrants. Several campsites are also available including the 'Sindibad' (❸ ☎ 044.84.75.53) downtown and the 'Tombouktou' (❸ ☎ 044.84.71.65) on the road to Mhamid, but both are very basic.

After Zagora, there is accommodation at Tamegroute ('Jnane Dar' ❸ ☎ 044.84.86.22), Tinfou ('Portes du Sahara' ❶ ☎ 044.84.70.02 and 'Repos des sables' ❸ ☎ 044.84.85.66, both poor value for money in 2002), Tagounite ('Bivouac Aït Isfoul' best avoided, 'Hotel la gazelle' ❸ ☎ 044.89.70.48 downtown, friendly and good value for money, and several camp sites among which the new 'Thé au Sahara' ❸ Fax 044.48.86.65 a few kilometres after the town at the pass of Jbel Bani – basic but built in a stunning landscape), Oulad Driss (a whole range of camp sites and inns, most of them possibly reasonable), and finally at Mhamid (several hotels, such as the 'Iriqi' ❸ ☎ and Fax 044.84.80.23) but the authors have not received a very warm welcome in this town and don't recommend staying there.

Some hotels also have berber tents for rent and others provide mattresses on the terrace; both solutions are of course cheaper and can be very pleasant when the weather is fine.

Strategy

This journey is best done in winter or early spring as temperatures can rapidly become uncomfortable from May onwards. Though it can be completed in one day starting from Ouarzazate (530km there and back), it is wise to allow at least two to three days to fully explore the region. A typical itinerary from Ouarzazate might be as follows: **Day 1:** Ouarzazate – Zagora; **Day 2:** Zagora – Mhamid – Zagora; **Day 3:** Zagora – Ouarzazate.

Most hotels offer camel rides (from an hour to several days), trekking, and four-wheel drive excursions. It is wise to book these through the major hotels as using smaller hotels or local 'guides' on the street might result in poor value excursions. Ali at 'Chez Ali' is a good contact, as are Ali Douini Brahim ('Caravane Mille Etoiles' 45 Avenue Mohamed V, Zagora, ☎ and Fax 044.84.62.35) for camel rides and Morhad ('Quad Evasion' Avenue Mohamed V, Zagora, ☎ 044.84.86.92) for four-wheel drive excursions.

The Draa valley can be divided into three distinct parts, each encompassing a different set of habitats; from Ouarzazate to Agdz, from Agdz to Zagora, and finally from Zagora to Mhamid.

From Ouarzazate to Agdz

The first part of the journey, from Ouarzazate to Agdz (70km), crosses the desolate mountainous, dry landscape of the Jbel Sarhro with its impressive cliffs and gorges. It peaks at the Tizi-n-Tinififft pass at 1660m. The birdlife is typical of rocky and bushy desert areas and includes such species as Desert and Thekla Larks, Desert, Black and White-crowned Black Wheatears, and Scrub Warbler. African Rock Martin has been reported breeding near the pass.

The few oases and gardens act as magnets for migrants; in March 2002, the cultivations and nearby desert close to the milestone 'Zagora 123' produced Egyptian Vulture, Rock Dove, European Bee-eater, Tawny and Tree Pipits, Common Bulbul, Nightingale, Redstart, Northern Wheatear, Desert and Black-eared Wheatears, Blackbird, Subalpine, Western Bonelli's and Willow Warblers, Chiffchaff, Woodchat Shrike (including one *badius*), House Sparrow, and House Bunting.

The first acacia trees (*Acacia raddiana*) appear just before reaching Agdz.

From Agdz to Zagora

Shortly after Agdz, the P31 starts following the Oued Draa, which is skirted by a series of superb palm-groves and gardens. Various vantage points give excellent views of the oued and several tracks cross the oued over low bridges. The following are just a few suggestions:

- Stop at the first bridge over the Draa 2km after Agdz, which leads to the 'Hotel Camping Jardin Tamnougalt'. This is a good spot to see Moroccan White Wagtail. While having a cup of mint tea at the hotel, the calls of Laughing Dove can be heard.
- Just before Tansikht (approximately 28km after Agdz) stop at a little dam on the Draa ('Barrage Tansikht'). From late March onwards, this is where the first Blue-cheeked Bee-eaters generally appear. Desert Warbler has also been seen here.
- 19km before reaching Zagora, stop at another small dam (signposted 'Barrage de dérivation d'Ifly' – see sketch page 120). The track runs through a wonderful palm-grove before reaching the oued: take the opportunity to visit this very special habitat.
- Enjoy breakfast at Café Essaada in downtown Zagora (Avenue Mohamed V) while watching Blue-cheeked Bee-eaters perched on the electric wires of the Royal Armed Forces, on the other side of the Avenue.

Species breeding in the palm-groves and gardens and along the Oued Draa include Barbary Partridge, Turtle, Laughing and Collared Doves, Little Owl, Hoopoe, Crested Lark, Common Bulbul, Cetti's and Western Olivaceous Warblers, Blackbird, Fulvous Babbler, Serin, Goldfinch. Blue-cheeked Bee-eater is fairly common all along the Draa valley and breeds as far north as Agdz. Stone-curlew is often heard calling at night.

The palm-groves along the oued also attract lots of migrants including Common Cuckoo, Scops Owl, European and Red-necked

Nightjars, Roller, Hoopoe, Wryneck, Tawny Pipit, Rufous Bush Robin Nightingale, Bluethroat, Redstart, Whinchat, many warblers, Pied Flycatcher, Woodchat Shrike, and Golden Oriole. Isabelline Wheatear has been recorded several times in March-April and is possibly a rare though regular spring migrant.

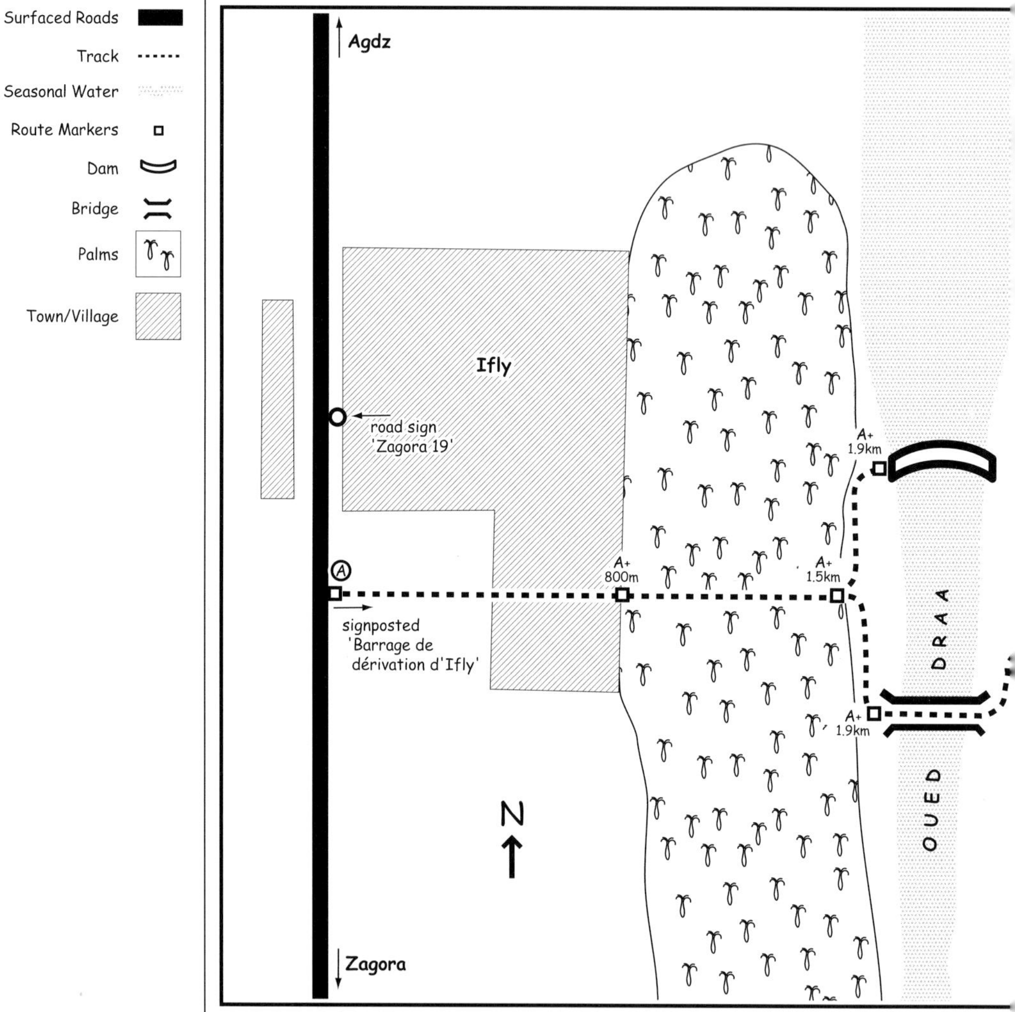

Several raptors also follow the Draa valley and Honey-buzzard, Black Kite, Egyptian Vulture, Short-toed and Booted Eagles, Marsh an Montagu's Harriers, and Osprey have all been recorded. Barbary Falc and Lanner are resident.

Many waterbirds have been recorded along the Oued Draa and the included Black-necked Grebe, Black-crowned Night-Heron, Ruddy Shelduck, Garganey, Marbled Duck, Water Rail and crakes, Common Crane, waders including Little Ringed Plover, Little and Temminck's Stints, Ruff, Jack Snipe, Redshank, Greenshank, and Green, Wood and Common Sandpipers, and even Gull-billed and Arctic Terns! Among the recorded rarities are Great Egret and Kittlitz's Plover.

Other interesting species recorded near Zagora have included Black-shouldered Kite, Tawny Eagle, and Arabian Bustard.

From Zagora to Mhamid

After Zagora and down to Mhamid, the road runs through the true Moroccan pre-desert. The bird density is generally low with White-crowned Black Wheatear often the most conspicuous bird, but other desert specialities have included Crowned Sandgrouse, Egyptian Nightjar, Scrub and Desert Warblers, Brown-necked Raven, and Desert Sparrow. In any case, the spectacular landscape makes the trip well worthwhile.

Some good spots include:

- The bridge over the Oued Draa downstream of the village of Tinfou. Stop here and walk along the oued and in the nearby bushes, tamarisks, and palm trees. Among the commoner species, Booted Eagle, Blue-cheeked Bee-eater, Moroccan White Wagtail, Fulvous Babbler, Western Olivaceous Warbler, and Brown-necked Raven have been recorded in late March.
- Some 50 kilometres after Zagora, the road climbs the Jbel Bani, a dramatic landscape, and then runs through a well preserved 'forest' of acacia trees. This is a possible spot for Lichtenstein's Sandgrouse.
- When there is water, the small 'Barrage Rgabi' (track sign-posted on the left side of the road before reaching Ouled Driss) is good for migrants. It is hard to believe but Black-crowned Night-Heron, Bar-tailed Godwit, and Green and Wood Sandpipers have been recorded here. Desert specialities have included Egyptian Nightjar, Scrub and Desert Warblers, and Brown-necked Raven.

Other Wildlife

Spanish Terrapin is common all along the Oued Draa. It is strange to see them drifting downstream but this makes it easier to understand why they are sometimes found in the most remote areas.

The West Sahara coast between Goulimine and Layoun

(11 – West Saharan Morocco)

A visit to this remote part of southwestern Morocco involves a return journey of some 860km, but provides an opportunity to experience the very special, sometimes wonderful landscape of this coastal, Saharan semi-desert region and to obtain an understanding of its birdlife. The sites described in the following pages are only some of the key ones (and the most accessible) between Goulimine and Layoun, but include the famous Khnifiss lagoon and the mouth of the Chebeika. The area is still poorly known from an ornithological standpoint, and it is possible to find new and exciting locations within it. The authors would be pleased to receive any trip reports and details of bird sightings.

Location

Goulimine lies some 200km south of Agadir on the main P41 road, which goes to Mauritania along the Atlantic coast. Starting from Goulimine, and driving south, three main natural regions are crossed: the Lower Draa, Tarfaya, and the Saquiat Al-Hamra. Layoun is at 429km from Goulimine.

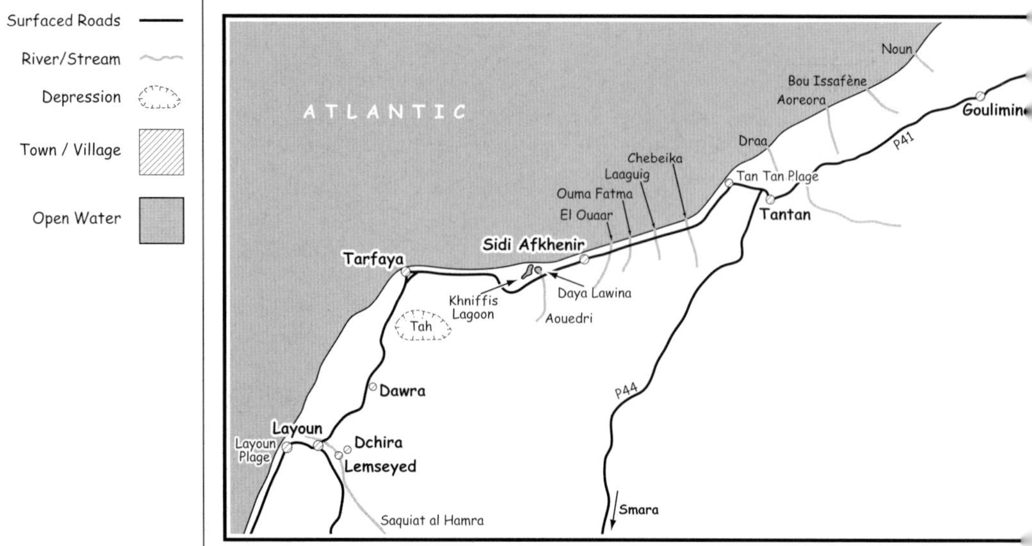

Accommodation

Hotels are few and far between in this region, though those prepare to camp or who are driving a camper van, will find many wonderful places to stop overnight. Tourism is not yet developed in West Sahara and so trips have to be planned accordingly.

Goulimine has several hotels in the basic or medium categories including the 'Bahich' (❸ 31 Avenue Abaynou. ☎ 048.77.21.78 – the best in 2002), and the noisy 'Salam' where several birders have stayed (❸ Route de Tantan. ☎ 048.87.20.57). Another alternative is to sleep at Fort Bou Jérif (see page 91) and start the drive south from there.

The wonder, recently built 'Hotel Maroc campement loisir' (❶/❸, Fort de Tafnidilt, Tantan, ☎ 066.84.81.65 and 065.11.20.92, e-mail: pisteaven@aol.com, http://perso.wanadoo.net.ma/pistard) is a perfec place to stay for a night or two near the Oued Draa. The owner, Guy Dubau, and his team can arrange visits to the mouth of the oued.

The next big city, Tantan, also has some hotels including the basic 'Bir Anzarane' (❸ 154 Boulevard Hassan II. ☎ 048.87.78.34, GSM 061.83.91.50, Fax 048.87.72.72) and the better 'Sables d'Or' (❸ Avenue Hassan II. ☎ 048.87.80.69, Fax 048.87.80.69 – the best in 2002) 20km later, the seaside resort of Tantan Plage (called 'El Ouatia' on the Michelin map) is a much better stop for birders, ideally located. It hol several basic hotel-restaurants including the 'Hôtels La Corniche' and 'Rose des Sables' (❷/❸. ☎ 048.87.93.73, GSM 068.80.03.13), 'Hôtel Café Restaurant La Belle Vue' (❸. ☎ 048.87.96.37, GSM 061.38.42.50 et 068.58.40.72, Faz 048.87.96.37). Tantan Plage may well develop quickly for mass tourism.

The next town, Tarfaya, is a kind of 'End of the World'; it fights har against the sand to survive and only just succeeds. It only has two ve basic hotels and these cannot be recommended. The only decent hote after Tantan Plage are thus at Layoun, some 285km later, but there is a very good and recommended campsite 'Luc et Martine' at Dawra, 72km south to Tarfaya where you can get accommodation under berb tent (❸ ☎ 048.99.45.46, www.geocities.com/leroibedouin).

Layoun city is developing quickly and holds several new hotels in the medium category, including the 'Mekka' (❷ Avenue Mekka. Layoune. ☎ 048.99.39.96, Fax 048.99.39.96) and the nearby 'Jodesa' (❸ 223 Avenue Mekka. BP 23. Layoune. ☎ 048.99.20.64, Fax 048.89.37.84). The more expensive modern 4* 'Sahara Line' (❶ Angle El Keraouane – 24 November. ☎ 048.99.54.54, Fax 048.99.01.55) is well run by Omar Erguibi and his friendly team, and some comfort might be appreciated after the long journey. Dinner is good value-for-money.

It is wise to buy water and food for lunch in the main cities as restaurants are few and far between on this route. If you are at Sidi Afkhenir at lunchtime stop in one of the small restaurant for fresh fish.

Strategy

Allow at least five days to cover the region. Starting from Goulimine, a typical 6-day journey can be as follows: **Day 1**: Goulimine – Tantan Plage, **Day 2**: Tantan Plage – coast up to the mouth of the Oued El Ouaar – Tantan Plage, **Day 3**: Tantan Plage – Khnifiss lagoon – Layoun, **Day 4**: quiet day: Layoun – Lemseyed and Dchira – Layoun, **Day 5**: Layoun – Tantan Plage, **Day 6**: Tantan Plage – Goulimine. Of course many other alternatives exist, especially for those camping and not restricted by the location of hotels.

A car is essential to fully explore the region. It is possible to fly to Layoun from Casablanca and hire a car there. The main P41 road which runs along the Atlantic is a good wide road but hundreds of lorries and coaches ply to and fro between the northern and southern cities and most of them drive at full speed (as is apparent from the results on the road sides). Just after the Oued El Ouaar (see below) petrol is tax free and the price is then halved; plan to refuel accordingly.

There are several police checkpoints along the route (3 in spring 2002: at the Oued Draa, near Ouma Fatma, and at the entrance of Layoun city) but passage is normally straightforward. Be sure to stop at the 'Stop' sign a hundred metres before the checkpoint and move forward only when asked. The region is safe and there are no mines along the main roads, but be sure to ask officials about any potential danger if intending to drive off-road to the interior.

The journey is best made during the migration periods in spring and autumn. Nevertheless, winter can be rewarding at some sites, such as the Khnifiss lagoon, and the period around Christmas and New Year is generally sunny and very pleasant – a major change from the north European climate at that time. Summers are hot in the interior but not so much so along the coast, however birds are then less numerous.

The sites and their birds

The 430 kilometres or so between Goulimine and Layoun are broken by a series of Saharan oueds and some other great sites that are described in sequence that follows. The first oued at the exit of Goulimine, the Oued Oum Laachar, is generally dry and is now almost entirely surrounded by the suburbs of the town. There is usually nothing of interest here but it is worth stopping some 5.5km later, at the Oued Sayed.

(**Note**: The distance to the various sites between Goulimine and Tantan Plage are given in relation to the bridge over the Oued Oum

Laachar and marked as 'G + xkm'. The distances to the sites between Tantan Plage and Layoun are given in relation to the 'crossroad of the dancing fishes' between the main road P41 and the road to Tantan Plage, and marked as 'TTP + xkm'. As they have been recorded using the odometer of a rented car, they may be approximate).

The Oued Sayed

(11a – Lower Draa) G +5.5km

The Oued Sayed rises in the Anti-Atlas and irrigates the palm-grove of Taghjicht (see 'Taghjicht' page 98); it flows into the Oued Noun som dozen kilometres downstream from this site (see 'Fort Bou Jérif and th Oued Noun' page 91).

Just after the bridge over the oued follow the track that heads to the west (right side of the road) for several hundred metres. There is generally water in this portion of the oued and Little Grebe, Moorhen, Common Bulbul, Rufous Bush Robin, Zitting Cisticola, Western Olivaceous Warbler, and Goldfinch breed; Marbled Duck and Nightingale possibly do so. The low bushes on the flat sandy area to the south of the oued are good for Scrub Warbler. Other species reported here include Barbary Falcon, Little Owl, Black-bellied Sandgrouse, Thekla Lark, Red-rumped and Desert Wheatears, and Fulvous Babbler.

Many migrants are attracted to the oued and interesting March-Apr records have included White Stork, Black Kite, Short-toed and Booted Eagles, Marsh and Montagu's Harriers, Black-winged Stilt, Collared Pratincole, Common Snipe, Green Sandpiper, Little Ringed Plover, European Bee-eater, Red-rumped Swallow, Bluethroat, Redstart, Sedge Reed, Melodious, Spectacled, Subalpine, Orphean, and Western Bonelli's Warblers, Pied Flycatcher and Woodchat Shrike.

Other interesting species recorded in autumn/winter have included Black-necked Grebe, Black-shouldered Kite, Tawny Eagle, Short-eared Owl, and Tristram's Warbler.

The Oued Boukila (G+11.7km) is generally dry – Scrub Warbler has been reported – and the sandy plains some 25km later are the next good area.

The sandy plains

(11a – Lower Draa) G +30-35km

Between G+30km and G+35km the P41 passes across vast sandy plains where locals try to cultivate poor cereal fields. This area is good for larks and Hoopoe, Desert and Bar-tailed Desert, Thekla, Thick-billed, Temminck's Horned Lark, Short-toed and Lesser Short-toed La have all been reported, sometimes in good numbers (especially Desert and Thick-billed) or even in large troops (Lesser Short-toed).

Other breeding species include Cream-coloured Courser, Black-bellied, Crowned and Spotted Sandgrouses, Red-rumped and Desert Wheatears, Desert Warbler, and Trumpeter Finch. The area is good for raptors (check the pylons of the electric line that follows the road) and in March-April Long-legged Buzzard and Lanner as well as migrating Black Kites, Montagu's Harriers and Booted Eagles have been seen. Golden, Bonelli's and Tawny Eagles have been recorded, especially in winter (with up to six of the latter in January 1990).

Short-eared Owl, Blue-cheeked Bee-eater, Tawny Pipit, and Tristram's Warbler are just some of the many species that have been recorded in late winter and spring.

The Oued Bou Issafène

(11a – Lower Draa) G+37-40km

The P41 follows the Bou Issafène for some three kilometres. During the migration periods, look in the halophytic vegetation which attracts migrants (for example Wryneck, Tawny Pipit, Nightingale, Rufous Bush Robin, Bluethroat, Black-eared Wheatear, Whitethroat, Spectacled, Subalpine and Orphean Warblers, Woodchat Shrike and Ortolan Bunting in spring). Migrant raptors have included Griffon Vulture, Short-toed and Booted Eagles, and Montagu's Harrier.

Residents include Cream-coloured Courser, Rock Dove, Thekla Lark, and Black Wheatear. This is a good place for Scrub Warbler as well.

The next good site is the Oued Draa (G+111km) but larks, wheatears, raptors and other species are likely to be seen on the drive and may call for roadside stops. In 2002, it was difficult to reach the shoreline with a regular car and a four-wheel drive was recommended.

The Oued Draa

(11a – Lower Draa) G +111km

Except in the driest months, there is generally water or at least some small ponds in the Lower Draa and these attract many birds, migrants as well as residents. The inner river bed is fringed by many tamarisks and glassworts while there are large bushes and less trees in the wide, sandy outer river bed. For many years it has been impossible to stop at the bridge because of police and military checkpoints (this might change in the future), and so it is better to drive for one or two kilometres on the track that runs northwest on the northern side of the oued and stop far enough from the road to birdwatch quietly. This track is marked as the 7092 on the Michelin map but its position is not very accurate (it is closer to the oued); it leads to an old French Foreign Legion fort called 'Tafnidilt' and to a newly-built, wonderful hotel, and then on to the mouth of the Oued, but large accumulations of sand made a four-wheel drive essential to get to the mouth in 2002.

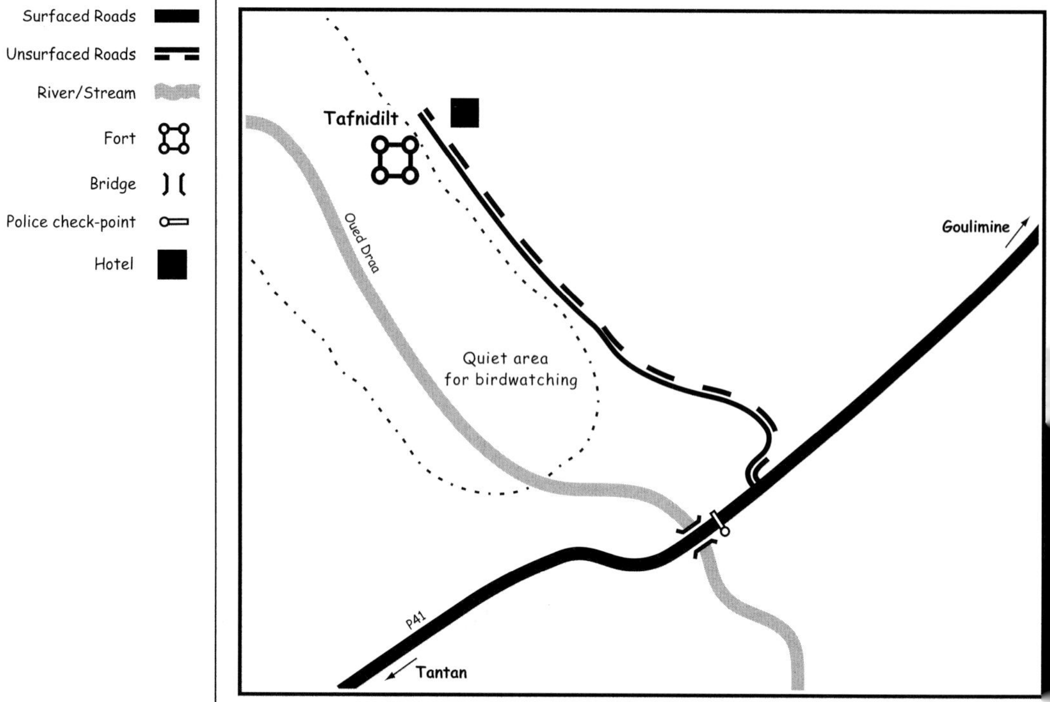

Breeding birds include Thick-billed Lark, and Desert and Red-rumped Wheatears in the sandy areas. Rufous Bush Robin and Western Olivaceous Warbler probably breed in the tamarisks along the oued but this ‘riparian woodland’ is especially good for migrant passerines (for example, many Turtle Doves, Barn Swallows, Sand Martins, Tree Pipits, Yellow Wagtails, Rufous Bush Robins, Northern Wheatears, Reed, Subalpine and Willow Warblers, and Woodchat Shrikes were present in late-April 2001).

Other records have included several waterbirds, including Moorhen (breeding), Black-winged Stilt and Common Sandpiper, and other specialities such as Tawny Eagle, Cream-coloured Courser, Stone-curlew, and Black and Mourning Wheatears.

Thick-billed Lark

The region is famous for mammals which include Jackal, Common Red Fox, Saharan Striped-Weasel, Striped Hyena, African Wild Cat, Wild Boar and two endemics (Tarfaya's Shrew and *Gerbillus occiduus*). 24 species of Amphibians and Reptiles have been recorded.

The entrance of Tantan city (G+129km) is marked by a very special Triumphant Arch ('The kissing Camels') but except this there is not much to be seen in the town (Collared Doves which arrived here in 1992, House Sparrows, and House Buntings) and the next good point is at Tantan Plage, some 20km later, which is on the sea.

Tantan Plage
(11b – Tarfaya)

This is a very convenient place to stay (see 'Accommodation' above) and also a good birdwatching spot, especially for seabirds. Half-an-hour on the terrace of the restaurant 'Equinoxe' on 25 April 2001 produced Northern Gannet, Moroccan race of Great Cormorant, Sanderling, Curlew Sandpiper, Dunlin, Whimbrel, Lesser Black-backed Gull, Gull-billed and Caspian Terns, Turtle Dove, Tree Pipit, White and Yellow Wagtails and Whinchat; as well as a spectacular group of 70 Spoonbills slowly flying north along the shore. Two Redwings and a melanistic Collared Dove were present in the gardens nearby.

Other records have included Sooty and Cory's Shearwaters, Black Stork, Lesser Crested Tern, Lanner and Razorbill. Plain Swifts have been claimed here (February 1992 and January 1997), and rarities have included Herring Gull (December 1993), Arctic Tern (March 1989) and Isabelline Wheatear (February 1997). Large numbers of Northern Gannets (1000+ in two hours in late-March 1989) and of Lesser Black-backed Gulls (14,000 in December 1993 and 30,000 in December 1995) have been recorded.

The busy fishing harbour is also worth visiting for gulls and terns. Ask for permission to enter from the police at the entrance. The jetty seemed to be promising for seawatching.

Going south from Tantan Plage, the next interesting feature is the sea cliffs.

The sea cliffs of Tarfaya region
(11b – Tarfaya)

Some 10km from Tantan Plage the first sea cliffs are reached and the road follows them for dozens of kilometres. They are very good for four breeding species: Moroccan race of Great Cormorant, Barbary Falcon, Lanner, and Yellow-legged Gull. All of these birds are fairly easy to find; in particular, Lanners and Barbary Falcons are widespread along the shore and sometimes provide spectacular demonstrations.

Seawatching is easy from the top of the cliffs but can be hampered by sea winds.

The next good sites are a series of three mouths of Saharan oueds, namely the Chebeika, Ouma Fatma and El Ouaar. At the beginning of the 2000s, the bridges over these oueds were still regarded as 'sensitive areas' by the Moroccan Authorities and stopping on these bridges was forbidden (this might well change in the near future as tourism develops); there was a police check point on the southern shore of the Chebeika.

The mouth of the Chebeika

(11b – Tarfaya) TTP + 30km

This tidal estuary holds a good variety of habitats with a sandy beach, mudflats, salt marshes, and tamarisks. Two lagoons, one on each side of the P41, are excellent for waterbirds. The Chebeika is certainly the nicest of the three oueds.

Stop on the descent to the oued, on the northern shore before crossing the bridge (there is a police checkpoint on the southern side) and walk along the edge of the lagoons or, better, on the top of the cliffs which offer good vantage points.

The next oued, the **Laaguig** (TTP + 43km) is dry most of the time.

The mouth of the Ouma Fatma

(11b – Tarfaya) TTP + 59km

The site covers approximately 150 hectares. Take the track that starts on the right half way up the climb on the southern side and park on top of the sea cliffs (campers are often present here as this is a famous fishing point). Walk along the sea cliffs to good vantage points. A well-preserved fossil of an Ichtyosaurus (some two metres long) lies at the bottom of the cliff, several dozen metres from the mouth on the southern shore.

The mouth of the El Ouaar

(11b – Tarfaya) TTP + 70km

This oued is also called Ez Zehar on the Michelin map. This is the smallest estuary of the three (it covers some 90 hectares) and has a good range of habitat including a sandy beach, stagnant salt or brackish waters with flooded vascular vegetation, and salt marshes. Two 'gueltas' (lagoons) are separated by the embankment of the main road.

The birdlife at these three estuaries is often rich. Large waterbirds have included Great Cormorant, Little Egret, Grey Heron, Black-crowned Night-Heron, Black Stork, Spoonbill and Greater Flamingo. Winter visitors include Little, Crested and Black-necked Grebes, Glossy Ibis, Shelduck and Ruddy Shelduck, Common Teal, Wigeon, Northern Shoveler, Mallard, Pochard, Tufted Duck, Common Scoter (2,000 in

January 1975 and 5,200 in January 1995 off the Chebeika), and Razorbill. A wide range of waders has been recorded, especially during migration: Collared Pratincole, Black-winged Stilt, Avocet, Oystercatcher (maximum 450 in January 1975), Ringed, Little Ringed, Kentish, Golden and Grey Plovers, Sanderling, Temminck's Stint, Dunlin, Curlew and Common Sandpipers, Common Snipe, Curlew, Whimbrel, Bar-tailed Godwit, Redshank and Spotted Redshank, and Turnstone. The Chebeika used to be a good wintering place for Slender-billed Curlew (123 in December 1974).

The three mouths are especially good for gulls and terns. Slender-billed, Audouin's, Lesser Black-backed, Yellow-legged and Black-headed Gulls, Gull-billed, Caspian, Royal (for example 25 at Ouma Fatma in August 1998 and 120 at El Ouaar in August 1995), Sandwich (3,100 at El Ouaar and 2,000 at Ouma Fatma in October 1998), Lesser Crested, Common, Little, Whiskered, and Black Terns have all been commonly recorded.

Kentish Plover breeds on the sandy beaches. Scrub Warbler is common and prefers the desert and halophytic plants. A small colony of Shags was discovered at the east of the mouth of the Chebeika at the beginning of the 1960s; breeding of Ruddy Shelduck and Common Raven have also been reported.

The cliffs and the 'regs' (stone desert) nearby are the habitat of Long-legged Buzzard, Lanner, Barbary Falcon, Cream-coloured Courser, Stone-curlew, Barn, Little and 'Pharaoh' Eagle Owls, sandgrouse (coming to drink at the oueds – especially Spotted and Crowned Black-bellied is at its southern limit at the Chebeika), Black and Desert Wheatears, and Trumpeter Finch. Golden Eagle and African Rock Martin have been found breeding some 15km upstream the Oued El Ouaar in 1961.

Moroccan rarities at the Chebeika have included Leach's Storm-petrel (November 1982), Little Shearwater (September 1971), Great Egret (April 1974 and December 1991), Greater Scaup (February 1987), Red Kite (December 1994), Great Black-backed Gull (December 2000), Roseate Tern (September 1971 and May 1997) and Arctic Tern (75 in December 1974) and Lesser Whitethroat (September 1971). Rarities at El Ouaar: Wilson's Storm-petrel (October 1982), Ring-necked Duck (January 1988) and Plain Swift (April 1997). There were 17 and 14 Ruddy Ducks at the mouths of the Ouma Fatma and El Ouaar in June 1986.

There are several petrol stations just after the El Ouaar which sell tax free petrol.

Driving further southwest along the P41, stop near the Moroccan Navy Control Station ('Poste de Contrôle de la Marine Royale') at TTP+85km to look at the impressive abyss open in the coastal cliff, where Rock Doves breed (the abyss is circled by just a few stones so be careful!).

Next is the village of Sidi Afkhenir (TTP+89km) where food is available and many small restaurants serve fresh fish. Those wishing to explore the next site, Khnifiss lagoon, have to apply for a special permit from the authorities here. This can be difficult and take a lot of time – especially as it can be hard to find the right person to issue it. Some birders have visited the site for a few hours without a permit, but they may have been lucky not to get caught and this cannot be recommended; nevertheless, a stop near the warden's house, a good vantage point, is generally no problem (see below). The row of coastal cliffs that started near Tantan Plage stops here and are replaced by sandy beaches; the first large sand dunes appear around TTP+105km.

Khnifiss lagoon

(11b – Tarfaya) TTP+114km

Khnifiss lagoon (also called Naïla, or Puerto Cansado, or Foum Agoutir on the Michelin map) covers some 6,000 hectares and is one of the most famous Moroccan sites along the Atlantic, set aside as a Natural Reserve in 1962 and as a Permanent Biological Reserve in 1983. This is also a Ramsar site. It is really a wonderful place that should not be missed.

To get to the eastern edge of the lagoon, take the small tarmac road signposted 'Naïla' that starts on the west side of the P41 at TTP+114km. After approximately 3km, park near the warden's house, on top of the cliffs that overlook the lagoon. Very good views can be had from here and the warden may give permission for you to take a short walk near the mouth of the lagoon ('Foum Agoutir'). The location of the birds depends on the tide but the northern half of the lagoon (closer to the mouth) is generally better.

Breeding birds include Slender-billed Gull (19 pairs in 1986), Yellow-legged Gull, Common Tern and Yellow Wagtail at the lagoon; Spoonbill, Greater Flamingo and Marbled Duck are possibly occasional breeders. Ruddy Shelduck, Black-winged Stilt, Moorhen and Eurasian Coot have been recorded breeding at the small Guelta el Aouina (a temporary lake fed by the Oued Aouedri), behind the warden's house. Great Cormorants are intermediate between the northern race *maroccanus* (white throat only) and the southern race *lucidus* (belly fully white); they breed (15-20 pairs in the 1980s) on the coastal cliffs and can be common at the lagoon between September and May (for example 300 in late March 1991). Other coastal cliff breeders include the vary rare Shag (bred in the 1960s) and Peregrine (*atlantis* form). Scrub Warbler is common in the bushes.

Regular passage migrants and/or winter visitors include grebes (Little and Black-necked), herons (Cattle and Little Egrets, Grey Heron), Spoonbill (160 in January 1995), Greater Flamingo (more than 1,000 in December 1985 and January 1995), Purple Heron, and Black-crowned Night-Heron. White and Black Storks are rarer. A few geese and ducks winter or pass through on migration and these have included Shelduck, Wigeon, Pintail, Northern Shoveler, Garganey, and Pochard.

The numbers of waders present are lowest in the summer (1,656 in June 1985) and highest in the winter (13,440 in December 1985). Dunlin, Bar-tailed Godwit, Red Knot and Oystercatcher are the four most abundant species, but a wide range of others have been recorded, including Black-winged Stilt, Avocet, Ringed, Little Ringed, Kentish and Grey Plovers, Turnstone, Curlew and Common Sandpipers, Little and Temminck's Stints, Sanderling, Redshank, Spotted Redshank, Greenshank, Ruff, Curlew, Whimbrel, Black-tailed Godwit, Common Snipe, and Collared Pratincole.

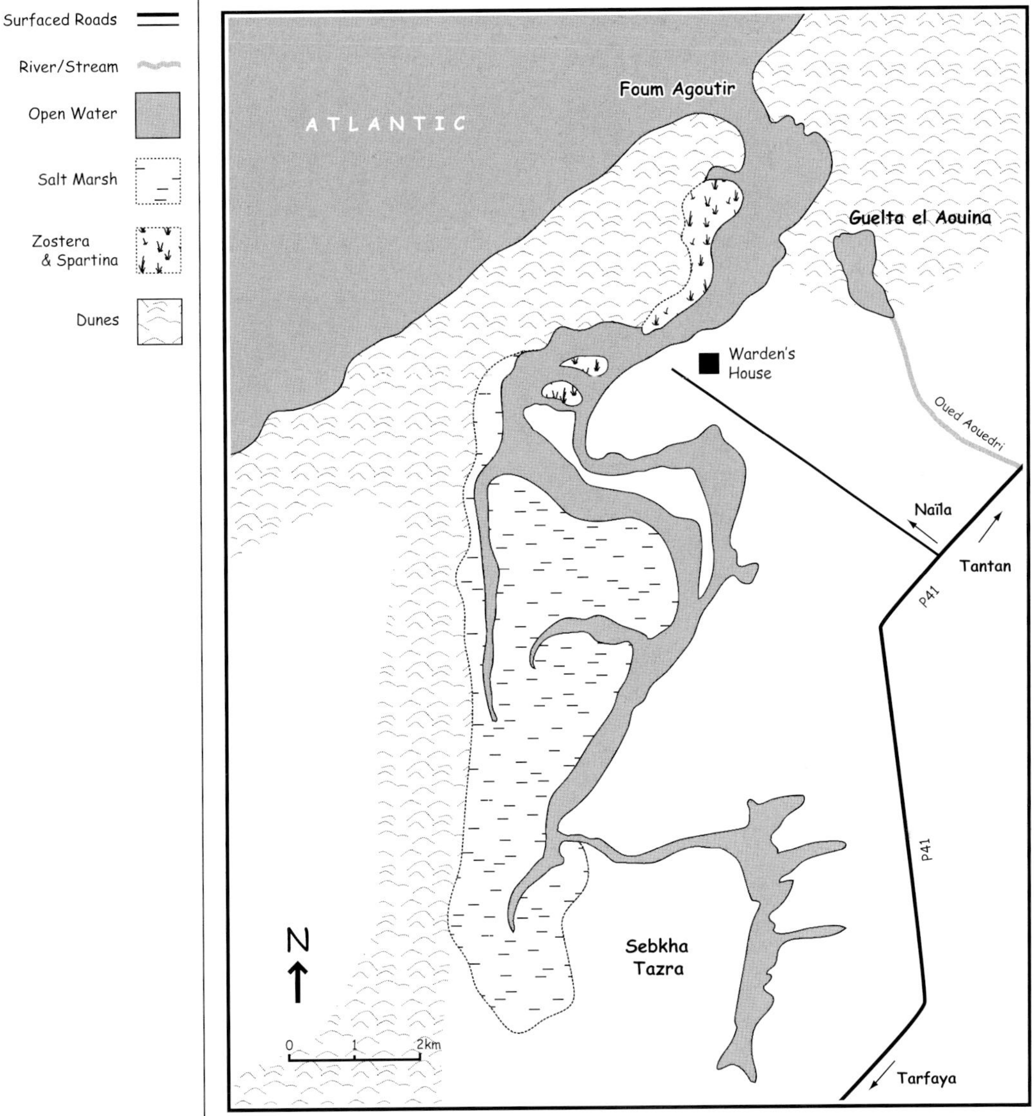

The numbers of gulls and terns also vary widely (5,400+ in October 1985) and the following have been recorded: Audouin's (sometimes more than 700 in winter), Slender-billed, Yellow-legged, Lesser Black-backed (5,000 in September 1972) and Black-headed Gulls, Gull-billed, Caspian, Royal (120 in August 1985), Lesser Crested, Sandwich (several thousand in September 1999), Common, Little, Black, and Whiskered Terns.

Here also, as at all the other sites along the West Sahara coast, the migration of passerines and other land birds can be impressive and Cuckoo, swifts, European Bee-eater, Hoopoe, Wryneck, Tree and Meadow Pipits, White Wagtail, Nightingale, Black and Moussier's

Redstarts, Northern Wheatear, Subalpine, Sardinian, Garden, Western Bonelli's and Willow Warblers, Chiffchaff, Pied Flycatcher, and Woodchat Shrike are just some of those that have been recorded.

The following have been recorded at the site, but are only occasional visitors: Great Crested Grebe, Greylag Goose, Gadwall, Common Teal, Mallard, Hen Harrier, Sparrowhawk, Bonelli's Eagle, Eleonora's Falcon, Red-knobbed Coot, Golden Plover, Northern Lapwing, Marsh Sandpiper, Kittiwake, Kingfisher, Roller, Skylark, Red-throated Pipit, Grey Wagtail, Robin, Ring Ouzel, Song Thrush, Redwing, Zitting Cisticola, Common Raven, Starling, Chaffinch, Goldfinch, and Linnet.

Rarities have included Wilson's Storm-petrel (October 1982), Great Egret (November 1985, December 1991), White-fronted Goose (April 1974), Brent Goose (May 1994), Greater Scaup (January-February 1986), Dark Chanting Goshawk (August 1972 and March 1991), Broad-billed Sandpiper (July 1999), Spotted Sandpiper (April 1990), Long-tailed Skua (April 1985), Herring Gull (April 1971 and September 1972), Great Black-backed Gull (December 1985 to April 1986), Roseate Tern (August-September 1972), Arctic Tern (September 1972), Short-eared Owl (December 1991 and 2001 and January 1995), Richard's Pipit (March 1991), Rock Pipit (March 1991), and Red-breasted Flycatcher (November 1985). The Khnifiss lagoon used to be an excellent wintering area for Slender-billed Curlew: 500-800 were present at the site in January 1964!

The road to Tarfaya
(11b – Tarfaya)

To the south of Khnifiss lagoon, the Sebkha (flat salty area) Tazra is generally devoid of water; the salt is collected by the local people (nice views of the salt pans some 10km after the crossroads of the P41 and the Naïla road to Tarfaya).

The road to Tarfaya crosses nice landscapes with tall sand dunes called 'Hassi Fleiga' (but beware of frequent sand accumulation on the road) and then follows the Atlantic coast where many ships have been run aground. Look for Royal Terns on the sandy beaches (up to 78 in October 1998), as well as for sea turtles and sea mammals blown ashore (many records).

At the end of a dead end road sometimes blocked by sand, Tarfaya (the former 'Cape Juby') is not a pleasant city which is a shame as the Cape forms a strategic position for seabird migration along the Atlantic coast. Cory's and Manx Shearwaters, Northern Gannets, Pomarine and Arctic Skuas and many gulls and terns have been recorded here.

The road to Layoun
(11c – Saquiat Al-Hamra)

The road between Tarfaya and Layoun (99km) crosses unusual landscapes. The Sebkha Tah some 30km to the south of Tarfaya, is a wide, dry, salty area which lies some 52 metres below sea level; it is devoid of birds. The village of Tah was the former edge of the Spanish Sahara (good, strong tea available). The oases of Dawra is watered by the Oued Al Marmoutha (dry most of the time) but tamarisks have developed in its bed and they are good for migrant passerines.

Cream-coloured Courser, Hoopoe Lark, Scrub Warbler, Desert Wheatear, and Brown-necked Raven are among the typical residents.

Layoun city is reached just after crossing the large, often dry Oued Saquiat al Hamra ('the Red River').

A trip through the Saquiat al Hamra

(11c – Saquiat Al-Hamra)

This short journey of some 50km (tarmac throughout) gives a good idea of this very desolate region of the lower Saquiat al Hamra. Plan half a day, and visit outside of the hottest months. Starting from Layoun, follow the main road to Smara; after some 10km, turn left to 'Oasis Lamsid'. Lemseyed ('Lamsid') is a tiny village with palm trees, tamarisks and reeds on the western side of the Oued Saquiat al Hamra; a small everlasting spring feeds a pool which attracts lots of migrants in spring, which in turn had attracted a pair of Barbary Falcons in April 2001. Fulvous Babblers have been seen at the pool and Green Toads reach the southern limit of their range here.

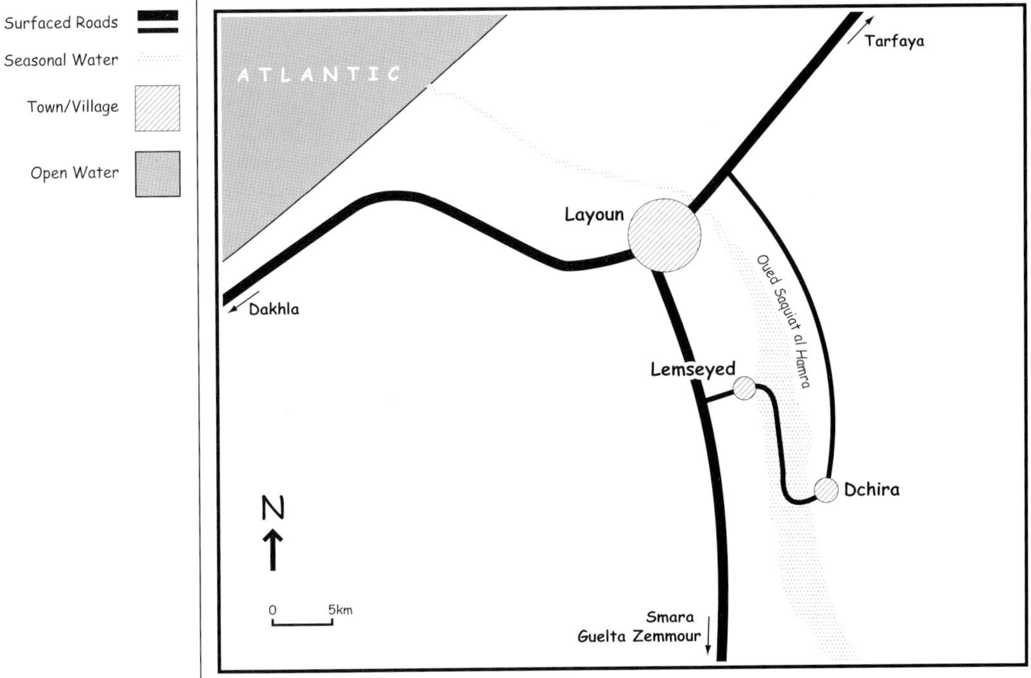

The road goes down to the bed of the oued, which is sprinkled with dry, burned tussocks against which the sand accumulates. Only a few birds live in this very harsh environment, among which is Temminck's Horned Lark. On the other (eastern) side of the Saquiat, the village of

Dchira is a stop for camel herds (caravans) and White-crowned Black Wheatear can be seen.

The small road then heads north on the plateau running along the Saquiat and finally reaches the main road P41 (26.5km). Turn left for Layoun. A good range of desert species can be seen on this drive including Long-legged Buzzard, Crowned and Spotted Sandgrouse, many larks (including Desert, Hoopoe, Short-toed), African Rock Martin, Desert and Red-rumped Wheatears, Scrub Warbler, and Trumpeter Finch. Spectacled Warbler possibly breeds here. Black-crowned Sparrow-Lark was seen here in April 1973.

Other wildlife

21 species of reptiles and amphibians were recorded between Tantan and Layoun in 1985-1986, and these include three toads (Mauritanian, Green and Brongersma's Toads), two sea turtles (Loggerhead and Leathery Turtles), Spur-thighed Tortoise, three geckos (Moorish, Helmeted and Brosset's Toed Geckos), Common Chameleon, two agamas (Bibron's and Spiny-tailed Agamas), Senegal Sand Skink, Golden Fringe-toed Lizard, four snakes (Montpellier, Schokar Sand, Diadem and Common Leaf-nosed Snakes), Egyptian Cobra, and two vipers (Lesser Cerastes and Puff Adder).

Dakhla Bay

(11d – Oued Ad-Deheb)

Dakhla Bay is the largest in Morocco and is a site of major importance for many migrant and wintering water birds. This really is the 'Deep South', the southernmost site described in this guide; it has recently been opened for tourism and is really worth visiting for waterbirds and for some of the least common species of the Western Palearctic.

Location

The town of Dakhla lies some 20km to the north of the Tropic of Cancer near the end of a narrow, 45km-long peninsula, almost at the southwestern tip of Morocco. This is basically a harbour where the main activities are related to fishing and fish processing.

The shallow bay has a maximum depth of only 22 metres; it is separated from the ocean by a sand and sandstone peninsula which is rather flat and peaks at only a few metres above sea level. The mouth is some 10km wide and the maximum width of the Bay is c.14km.

The quickest way to get there is by plane; Royal Air Maroc (http://www.royalairmaroc.com/) operates a regular service from Casablanca airport to Dakhla. An alternative solution is driving from Layoun (514km, see page 132); obviously this is a much longer, though pleasant, trip on a good road close to the Atlantic and this is possibly the most convenient solution for those intending to visit the sites between that town and Dakhla (several good hotels are available en route, at Boujdour).

Dakhla is the final Moroccan stop before crossing the last part of the desert to Mauritania (some 300km).

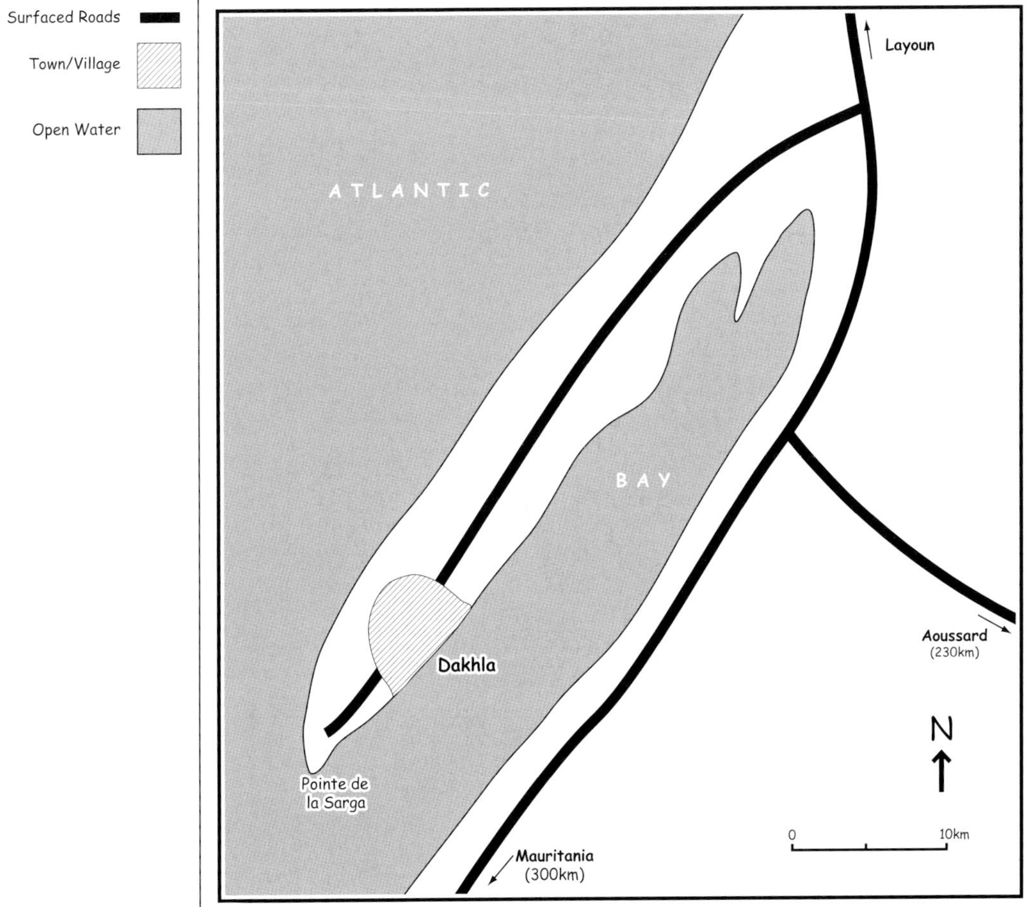

Accommodation

Dakhla holds a dozen or so very basic and very cheap (❸) hotels, e.g. 'Hôtel Bahia' (near the restaurant 'Casa Luis'. ☎ 067.64.57.13), 'Hôtel Café Restaurant Al Mouaouama' (corner Avenue Imlili and Avenue Al Moukaouama, n°15. ☎ 048.93.07.20), 'Hôtel Aigue (Avenue Sidi Ahmed Laaroussi near the restaurant 'Samarkand'. ☎ 048.89.73.95), 'Hôtel Al Massira' (near the fish market) and 'Hôtel du 14 Août (Avenue du Prince, near the airport). In 2002, there were only two hotels in the medium category, the friendly and good value for money 'Hôtel-Résidence Erraha (❷ Avenue Hamed Ben Chakroune, Massira II. In the southern part of the town near the great Mosque. ☎ 048.89.88.11, Fax 048.89.88.11. Our favourite in 2002) and the fallback 'Hôtel Doumss' (❷ Avenue El Walaa, quartier Moulay Rachid. B.P. 49. ☎ 048.89.80.46, 47 and 48, Fax 048.89.80.45). If money is no problem, the Hôtel Sahara Regency' (❶ Avenue El Walae. ☎ 048.89.80.70, 71, 72, Fax 048.89.70.45, e-mail: riadevasion@iam.net,ma, http//www.riad.ma) offers high standard accommodation. The campsite 'Camping Moussafir' is available at the entrance of the town. Two recommended places for breakfast and dinner are the restaurant 'Samarkand' and 'Casa Luis' near the waterside but many basic ones were also available (salads, chicken, fish, but also delicious camel kebabs).

It is essential to buy your picnic and water in Dakhla (at the central market downtown) before going out birding as there is nowhere else to buy supplies.

Strategy

The climate is fairly uniform at Dakhla and rain is exceptional; fresh water only comes from deep wells. Because of the trade winds, the temperature is nice or even cool during the day and cold at night. The white sand beaches offer good opportunities for bathing but the waters remain fairly cold until summer.

Most periods are excellent for birding; spring and autumn will be good for migrants and winter for many waders and gulls. Summer is probably the least interesting period. It is possible to travel all around the bay to look for the best resting and feeding places on beaches and mudflats (a telescope is essential); the largest intertidal areas are in the northern part of the Bay, and these often prove to be excellent. The 'Pointe de la Sarga' at the tip of the peninsula is also a good place for resting gulls and terns and to watch bird movements.

Allow one day for a trip to the interior on the road to Aoussard, to experience this very particular habitat (the real 'Deep South') and look for some of the rarest birds.

Don't forget that the region was at war with the Front Polisario only a few years ago and mines have been spread; it is essential to stay on the tarmac roads, above all in the interior.

Birds

The bay holds more than 21% of the total number of waders that winter in Morocco and several thousand generally winter here. 36,000 waders were counted in January 1995. The most abundant are Red Knot, Dunlin and Bar-tailed Godwit (7,000 to 9,000 birds). Oystercatcher, Ringed, Kentish and Grey Plovers and Sanderling are also common. Avocet, Black-winged Stilt, Golden Plover, Little Stint, Curlew and Common Sandpipers, Whimbrel, Curlew, Common Snipe and Jack Snipe, Marsh Sandpiper, Redshank, Spotted Redshank, Greenshank and Turnstone are less frequent or even rare.

In winter, several thousand gulls and terns are attracted by the fishing activities and these include Lesser Black-backed Gulls (up to 22,000 in Dec 1995) and Audouin's Gulls (up to 2,500 in the Bay in January 1995). Other regular species include Northern Gannet (1,500 in the Bay in mid-Jan 1989), Great Cormorant (747 in January 1995), Greater Flamingo (980 in January 1995), Slender-billed Gull, Caspian Tern and Grey Phalarope (off the coast). Little Egret, Grey Heron, Glossy Ibis, Pintail, Northern Shoveler, Pochard, Ferruginous Duck, Common Scoter, Marsh Harrier, Black-headed Gull, Kittiwake, Gull-billed, Arctic, Little and Black Terns, Short-eared Owl, and Red-throated Pipit are rarer or only occasional in winter.

The bay is a major stopping point for many waterbirds during their migration to and from Africa. Among the most interesting species are Spoonbill, Lesser Crested Tern (more than 2,000 recorded), Royal Tern (300+ in August 1998) and Sandwich Tern (up to 2,220 in October 1998). Many other migrants should be seen and Osprey, Wryneck,

Red-throated Pipit, and Melodious Warbler are only some of those that have been reported. Breeding birds include Great Cormorant, Kentish Plover, and Yellow-legged Gull.

Passages at sea can also be rewarding and day totals of more than 10,000 Cory's Shearwater have been counted in October. Great, Sooty and Manx Shearwaters, Leach's Storm-petrel, Pomarine, Arctic and Great Skuas and huge numbers of Northern Gannets have also been recorded.

Hoopoe and Thekla Larks, Desert, Red-rumped and Black Wheatears are among the commoner passerines. A trip in the interior should also produce species such as Cream-coloured Courser, Spotted Sandgrouse, Thick-billed, Temminck's Horned, Desert and Bar-tailed Desert Larks, Trumpeter Finch, Desert Sparrow and Brown-necked Raven and possibly the rare Houbara Bustard. Black-crowned Sparrow-Lark and Dunn's Lark have been recorded recently.

The first House Sparrow was reported in 1988 and the first Collared Dove in 1998; this species is now only some 360km from the closely related African Collared Dove which breeds in Mauritania. There are also some records of Laughing Doves but it is not sure which subspecies is involved, either the northern *phoenicophila* or the African *senegalensis*.

Accidental visitors have included White-faced and Madeiran Storm-petrels, Long-tailed Cormorant, Western Reef Egret, Red-necked Phalarope, Long-tailed Skua, Namaqua Dove and Isabelline Wheatear. Dakhla should also be one of the best Moroccan places to see some African specialities, perhaps including African Spoonbill, Lesser Flamingo, Grey-headed Gull or Bridled and Sooty Terns.

SELECTIVE BIRD LIST

Bald Ibis. Once much commoner, now restricted to a few coastal colonies in Haha near Tamri (page 57) and Souss in the National Park of Souss-Massa (page 81). Surveys in 1998 and 1999 located 60 breeding pairs (200-250 birds).

Ruddy Shelduck. A scarce to locally common resident; main nesting localities are in Western Middle Atlas, such as at Aguelmane Sidi Ali (see 'The region of the Lakes', page 54), and in desert areas south of the High Atlas, such as at Ouarzazate lake (page 113).

Marbled Duck. An uncommon bird, except in a few places such as Lac de Sidi Bourhaba (page 37) where it breeds and wintering numbers can reach several hundred. The mouth of the Oued Massa (page 84) is another good place to find it. Birds wander and can potentially be seen on most of the dayets and lakes described in this guide, including temporary lakes in desert areas such as the Dayet of Merzouga in the Tafilalt (page 106).

Ruddy Duck. Small numbers have been regularly observed since the early 1990s at three sites: Merja Halloufa and Merja Bargha (page 31), and in the Royal Reserve of Douyiet in Saïs.

White-headed Duck. Disappeared in the early part of the 20th century, and was considered an accidental visitor until the 1980s. Following population increase in southern Spain, sightings have subsequently increased in numbers, and now it is sometimes recorded at Merja Bargha (page 31) and at Douyiet (and sometimes also hybrid White-headed x Ruddy Duck).

Black-shouldered Kite. An uncommon local resident with a stable or slightly increasing population of a few hundred pairs. It is found in four distinct areas: the plains and hills of the northwest (including Zaë page 39), northeastern Morocco, the Marrakech region, and the Souss (page 88). Inhabits open ground, arable areas, and grassland with trees and hedges; also in palm-groves near Marrakech.

Dark Chanting Goshawk. A very scarce resident on the verge of extinction. Restricted to the plain of the Souss (page 88), in argana forest and olive-groves. Its biology in Morocco is almost unknown.

Tawny Eagle. A very rare resident breeder, with an estimated population of a few dozen pairs. Found in the past over a large part of Morocco, it is now scarce and probably restricted to the plain of the Souss (page 88) and the neighbouring areas of the High and Anti-Atlas. From time to time it is seen in other parts of the country, especially the Tafilalt (page 106), the mouth of the Oued Massa (page 84), and West Sahara, especially to the south of Goulimine (page 121).

Eleonora's Falcon. A very rare migrant breeder which arrives at its colony from late April and leaves in the second half of October. There is only one extant colony – the islands off Essaouira (675 breeding pairs in 2001) (page 49). The other colony off the coastal cliffs of Salé, near Rabat (21-25 pairs in 1981) has almost disappeared after the extension of the city boundaries.

Barbary Falcon. An uncommon resident throughout most of Morocco, but the precise distribution is still inadequately known because of possible confusion with Peregrine. Schollaert and Willem (2000) have summarized its status, following the records analysed by the Moroccan Rare Birds Committee.

Double-spurred Francolin. A very local resident, with most recent records in the Zaër (page 39) where the population has increased due to the creation of 'hunting reserves'. It inhabits cork oak woods with dense undergrowth, especially near small wadis.

Helmeted Guineafowl. Until 1920 the species was common in the Zaër, the Central Plateau, and the western foothills of the Middle Atlas. The population gradually decreased and the species has disappeared from Morocco. The last records of wild birds were in 1974.

Andalusian Hemipode. A very rare resident with secretive habits, whose biology is almost unknown in Morocco. Nowadays it appears to be restricted to the fallow fields of two regions, the eastern Mediterranean coast and the Atlantic plains between Larache (page 28) and Safi.

Purple Swamphen. A very local resident best looked for at the marshes of the Lower Loukkos (page 28). It has recently spread south and in the early 2000s colonized Merja Bargha (page 31), Merja Zerga (page 34), and Lac de Sidi Bourhaba (page 37). It has also re-colonized the Lower Moulouya marshes in Eastern Morocco.

Red-knobbed Coot. A local resident common at certain marshes such as the Lower Loukkos (page 28) and on certain lakes such as those of the Middle Atlas (page 54) or Lac de Sidi Bourhaba (page 37).

Demoiselle Crane. Formerly a very rare migrant breeder restricted to the Middle Atlas and the Central Plateau. No certain record since the mid-1980s.

Houbara Bustard. Previously a common resident on the eastern High Plateaux and in the southern deserts. Nowadays, because of over-hunting, especially by Middle Eastern falconers, the species has become very rare. Most recent records come from the eastern steppes, the Tafilalt (page 106), and Tagdilt (page 103).

Arabian Bustard. Previously common in uncultivated steppes of most of Morocco (race *lynesi*), but it has disappeared since the middle of the 20th century. Now the only records are from semi-desert areas south of the Atlas, especially in the Tafilalt near Merzouga (page 106), from September to April. Sightings probably involve the African race *stieberi*.

Great Bustard. A small relict population of a hundred birds still survives in the north, near Asilah (page 26) and Larache (page 28).

Slender-billed Curlew. Commonly reported as a wintering species until the 1960s in Tangier Peninsula and Rharb, at Sidi-Moussa and Oualidia lagoons (page 44), at the mouth of the Oued Massa (page 84), and along the Sahara coast (page 121). The largest numbers were in

West Saharan Morocco at Khnifiss Lagoon (page 130). Since then it has declined markedly. From the mid-1980s through the mid-1990's, only 2-5 birds were recorded each winter at Merja Zerga (page 34), and the last record was in February 1998. This species is now very close to global extinction and any sightings should be reported at once to BirdLife International in Cambridge, UK.

Audouin's Gull. Breeds only on the Mediterranean coast, where the largest colony is located on the Chafarines islands, off the mouth of Oued Moulouya (2,330 nests in 1989). It winters mainly along the Mediterranean and the West Sahara Atlantic coasts and is regularly seen at most of the other lagoons and wadis mouth during migration periods.

Royal Tern. An uncommon passage migrant along the Atlantic coast, from mid-March to mid-May and from August to October. A few oversummer or overwinter.

Lichtenstein's Sandgrouse. A resident strictly associated with the occurrence of *Acacia seyal* in the stone desert between the Jbel Bani and the Middle Draa, especially south of Foum Zguid and Tata (page 96).

Crowned Sandgrouse. A resident which inhabits all the desert areas south of a line from Goulimine in the southwest, through Ouarzazate and Merzouga to Figuig in the east. It favours rocky slopes with sparse vegetation and stony desert.

Spotted Sandgrouse. A resident whose breeding distribution in Morocco is very similar to that of the previous species. It is the commonest sandgrouse in Tafilalt (page 106). It favours sandy plains and open desert with sparse vegetation cover.

Black-bellied Sandgrouse. A resident which occurs in the eastern High Plateaux (up to the Mediterranean coast), the North and Mid-Atlantic Plains, the Haouz, the plain of the Souss (page 88), the southern foothills of the High Atlas, the Anti-Atlas and along the West Sahara coast down to the Oued Chebeika (page 128).

Pin-tailed Sandgrouse. A resident in the steppes of the eastern High Plateaux and the near-desert south of the High and Anti-Atlas. Like the Crowned and Spotted Sandgrouses, it wanders in relation to climatic conditions.

Collared Dove. First recorded in 1971 it then spread rapidly throughout the country and is now a common resident, even in desert areas. The colonisation process was described by Bergier *et al*. (1999).

Laughing Dove. After the first record in 1930, the species was found again in the palm-grove of Marrakech in 1982. Now this is a common species in most palm-groves of the Jbel Bani (page 96), in the Draa valley (page 116), and here and there including the gardens near Massa (page 84) and the Tafilalt (page 106). The colonization process has been detailed by Bergier *et al*. (1999).

Eagle Owl. Uncommon but widespread resident, including in desert areas. Two races *ascalaphus* and *desertorum* occur and are much smaller than the European races. These are sometimes split from Eurasian Eagle Owl and treated as a separate species called Pharaoh Eagle Owl *Bubo ascalaphus*. The status of the Iberian form of Eurasian Eagle Owl *B. b. hispanus* is unclear in Morocco.

Marsh Owl. A rare resident, restricted to certain places within the Tangier-Casablanca-Meknès triangle. This owl has suffered from the destruction and drainage of its marshy habitat. A few pairs still survive at Lac de Sidi Bourhaba (page 37) and a larger population occurs at Merja Zerga (page 34). Details have been published by Bergier & Thévenot (1991).

Red-necked Nightjar. A widespread but uncommon breeding migrant and rare winter visitor in non-desert Morocco, which favours open woodlands and coastal dunes. It is regular in the eucalyptus woods near the mouth of the Oued Souss (page 82).

Egyptian Nightjar. A scarce to locally common migrant breeder from the Central Anti-Atlas in the west to Saharan Atlas in the east, although most of the current records come from the Tafilalt (page 106). It is present from mid-March to October. It bred southwest of Marrakech in the 1960s.

Plain Swift. Small black swifts (often reported as Plain Swifts) have been recorded during the breeding season along the coastal cliffs south of Tamri (page 57) and in winter along the Atlantic coast, especially in the Souss at the mouth of the Massa (page 84). The exact status of this species is very unclear and all records should be sent to the Moroccan Rare Birds Committee.

White-rumped Swift. A very rare migrant breeder, probably restricted to a few valleys in the Central High Atlas. It has been recorded in the area of Imlil south of Marrakech (page 65), and in the mountains to the northwest of Ouarzazate. Almost nothing is known about its biology in Morocco.

Blue-cheeked Bee-eater. A migrant breeder, present only in the Figuig region in the far southeast, in the Tafilalt (page 106), and in the Draa valley (page 116), sometimes as far north as Ouarzazate (page 113). It arrives at the end of March and leaves before the end of September. It breeds in banks of wadis and in slopes of irrigation channels, sometimes alongside European Bee-eater. It is frequently seen on wires beside roads.

Levaillant's Green Woodpecker. An uncommon to locally common resident in the North, in the High Atlas, such as at Imlil (page 65), and in the Middle Atlas, such as in the region of the lakes (page 54).

Dupont's Lark. The least known lark in Morocco. It inhabits the steppes of the eastern High Plateaux and extends up the upper valley of the Oued Moulouya (page 77), where most of the recent sightings come from.

Thick-billed Lark. A resident which wanders outside the breeding season. This large lark occurs on stone desert and hammadas with sparse vegetation in two separate breeding areas: an eastern populatio from the eastern High Plateaux and the Upper Moulouya (page 77) through the Tafilalt (page 106), the Dadès-Draa (Tagdilt Track, page 10 and Ouarzazate, page 113) and to the Central Anti-Atlas, and a southwestern population in West Sahara (page 121). In autumn and winter, birds can gather in flocks of up to several dozen.

Temminck's Horned Lark. A common resident with the same distribution as the Thick-billed, but which favours habitats with more vegetation, such as *Artemisia* steppe and Halfa grass in the Upper Moulouya (page 77).

African Rock Martin. Very similar to the Moroccan race of Crag Marti and thus confusion is frequent. Seems to be restricted to the Saharan regions and the only proven breeding records come from the Tarfaya region. Detailed records are much needed and should be sent to the Moroccan Rare Birds Committee.

Isabelline Wheatear. A rare but probably regular spring passage migrant in East Saharan Morocco, especially in Tafilalt (the region of Er-Rachidia, page 74, and Erfoud, page 106) and in Dadès-Draa (Ouarzazate, page 113, and Tagdilt track, page 103). Most records are from February to April.

Mourning Wheatear. A scarce resident of broken rocky areas of the southern slopes of the Eastern High Atlas, such as at Barrage Hassan-Addakhil (page 74), and of the Central High Atlas, such as near Amerzgane, north of Ouarzazate (page 113).

Scrub Warbler. An uncommon resident, patchily distributed, mainly in the steppes of semi-desert areas south of the High Atlas. Best looke for in the Tafilalt (page 106) and above all in the Lower Draa, from Goulimine to Tantan (page 121).

Tristram's Warbler. An altitudinal migrant which only breeds in the scrubs of the High Atlas between 1,000m and 2,800m. South of the Atlas, it is only a winter visitor to Dadès-Draa (near Ouarzazate, page 113), Souss (page 88), Western Anti-Atlas and Lower Draa (near Goulimine, page 121). It is common from December to February in the tamarisks near Merzouga (page 106).

Desert Warbler. Restricted to sandy soils in desert areas and generall considered a scarce resident, but mainly seen from January to June in the Tafilalt (page 106).

Fulvous Babbler. A fairly common resident in all the semi-desert anc desert areas south of the High Atlas, from the Figuig in the southeast the West Sahara, through the Tafilalt (page 106), the Draa valley (pag 116), the Jbel Bani (page 96), the plain of the Souss (page 88), and the West Sahara (page 121). It inhabits open land with scattered bushes and shrubs and open palm-groves. The species is gregarious and is generally encountered in groups of up to 10-15 birds.

Black-crowned Tchagra. An uncommon resident, known from Tangi east along the Mediterranean coast and up into the Rif, the Zaër (page 39), the Central Plateau, the area around Essaouira (page 49), the Haha (Tamri, page 57), the plain of the Souss (page 88) and the nearby coast (mouth of Oueds Souss and Massa, page 82-84), and the Lower Draa. It lives in scrub, and in spring it is most easily located by its voice.

Brown-necked Raven. An uncommon to locally common resident widespread in West Sahara north to Goulimine (page 121), in Middle Draa, such as in the oases of Bani (page 96), and in Tafilalt (page 106) where numbers have recently increased near Merzouga.

Desert Sparrow. An uncommon resident, which reaches one of its northwest limits in the Tafilalt (page 106) where it occurs around the Erg Chebbi.

Crimson-winged Finch. A rare resident confined to high altitudes in the Middle Atlas, such as Jbel Bou Iblane, and the High Atlas. One of the best places to observe the species is around Oukaïmeden (page 67), where it can be seen from 2,300m to 3,000m in the breeding season, and down to 1,400m in winter. Also sometimes recorded in the gorges of the Oued Todra (page 71) and in winter at Tagdilt (page 103).

FULL SPECIES LIST

After Thévenot, Vernon and Bergier 2003, adapted

This list includes the 452 wild species that have been positively identified at least once in Morocco. It does not take into account the unsuccessfully introduced species or the species whose sight records have not been documented enough to be accepted. A code gives the present status of each species. The species followed by an asterisk are those which have to be submitted to the Moroccan Rare Birds Committee. Please send any such records to:

Prof. Jacques Franchimont
Quartier Abbass Lemsaadi,
rue n°6, n°22
50 000 Meknès V.N.
Maroc
j.franchimont@iam.net.ma

or

Patrick Bergier
4 Avenue Folco de Baroncelli
13210 – Saint Rémy de Provence
France
pbergier@yahoo.fr

Key to checklist

The status is abbreviated as follows:

rb	resident breeder
ob	occasional breeder
fb	former breeder, disappeared
mb	migrant breeder, present only in breeding period
pm	passage migrant
wv	regular winter visitor
ow	occasional winter visitor
as	accidental species
(i)	introduced species
?	indicates some doubt over the status

The scale gives a general indication of the abundance of each species its normal habitat

1	very common
2	common
3	uncommon
4	scarce
5	rare or very localised

fb,as?	☐	Ostrich (*Struthio camelus*)*
as	☐	Red-throated Diver (*Gavia stellata*)*
as	☐	Black-throated Diver (*Gavia arctica*)*
as	☐	Great Northern Diver (*Gavia immer*)*
rb2wv2	☐	Little Grebe (*Tachybaptus ruficollis*)
rb3wv3	☐	Great Crested Grebe (*Podiceps cristatus*)
as	☐	Slavonian Grebe (*Podiceps auritus*)*
rb4wv3	☐	Black-necked Grebe (*Podiceps nigricollis*)
as	☐	Black-browed Albatross (*Diomedea melanophris*)*
as	☐	Fulmar (*Fulmarus glacialis*)*
as	☐	Bulwer's Petrel (*Bulweria bulwerii*)*
mb5pm1ow	☐	Cory's Shearwater (*Calonectris diomedea*)
pm4	☐	Great Shearwater (*Puffinus gravis*)*

pm4ow	☐	Sooty Shearwater (*Puffinus griseus*)
pm3ow	☐	Manx Shearwater (*Puffinus puffinus*)*
pm2wv3	☐	Balearic Shearwater (*Puffinus mauretanicus*)
pm2wv5	☐	Yelkouan Shearwater (*Puffinus yelkouan*)
as?	☐	Little Shearwater (*Puffinus assimilis*)*
pm4ow	☐	Wilson's Storm-petrel (*Oceanites oceanicus*)*
as	☐	White-faced Storm-petrel (*Pelagodroma marina*)*
pm3wv4mb?	☐	European Storm-petrel (*Hydrobates pelagicus*)
pm3wv3	☐	Leach's Storm-petrel (*Oceanodroma leucorhoa*)*
as	☐	Madeiran Storm-petrel (*Oceanodroma castro*)*
as	☐	Brown Booby (*Sula leucogaster*)*
pm1wv1	☐	Northern Gannet (*Morus bassanus*)
as	☐	Cape Gannet (*Morus capensis*)*
rb3wv2	☐	Great Cormorant (*Phalacrocorax carbo*)
rb5wv4	☐	Shag (*Phalacrocorax aristotelis*)
as	☐	Long-tailed Cormorant (*Phalacrocorax africanus*)*
as	☐	African Darter (*Anhinga rufa*)*
as	☐	White Pelican (*Pelecanus onocrotalus*)*
b?fbpm4wv5	☐	Bittern (*Botaurus stellaris*)*
mb5pm3ow	☐	Little Bittern (*Ixobrychus minutus*)
mb3pm2ow	☐	Black-crowned Night-Heron (*Nycticorax nycticorax*)
mb4pm3ow	☐	Squacco Heron (*Ardeola ralloides*)
rb1pmwv	☐	Cattle Egret (*Bubulcus ibis*)
as	☐	Western Reef Egret (*Egretta gularis*)*
rb2pm2wv2	☐	Little Egret (*Egretta garzetta*)
pm5ow	☐	Great Egret (*Ardea alba*)*
pm2wv2ob	☐	Grey Heron (*Ardea cinerea*)
pm3mb5ow	☐	Purple Heron (*Ardea purpurea*)
as	☐	Yellow-billed Stork (*Mycteria ibis*)*
pm4ow	☐	Black Stork (*Ciconia nigra*)
mb2pm2wv5	☐	White Stork (*Ciconia ciconia*)
pm4wv4ob	☐	Glossy Ibis (*Plegadis falcinellus*)
rb5	☐	Bald Ibis (*Geronticus eremita*)
pm3wv4ob	☐	Spoonbill (*Platalea leucorodia*)
pm2wv2fb	☐	Greater Flamingo (*Phoenicopterus ruber*)
as	☐	Lesser Flamingo (*Phoenicopterus minor*)*
as	☐	Fulvous Whistling-Duck (*Dendrocygna bicolor*)*
as	☐	Mute Swan (*Cygnus olor*)*
as	☐	Whooper Swan (*Cygnus cygnus*)*
as	☐	Bean Goose (*Anser fabalis*)*
as	☐	White-fronted Goose (*Anser albifrons*)*
wv4	☐	Greylag Goose (*Anser anser*)
as	☐	Snow Goose (*Anser caerulescens*)*
as	☐	Barnacle Goose (*Branta leucopsis*)*
as	☐	Brent Goose (*Branta bernicla*)*
rb3	☐	Ruddy Shelduck (*Tadorna ferruginea*)
wv3	☐	Shelduck (*Tadorna tadorna*)
as	☐	Spur-winged Goose (*Plectropterus gambensis*)*
as(i)	☐	Mandarin Duck (*Aix galericulata*)*
wv1	☐	Wigeon (*Anas penelope*)

as	☐	American Wigeon (*Anas americana*)*
wv3ob	☐	Gadwall (*Anas strepera*)
wv2	☐	Common Teal (*Anas crecca*)
as	☐	Green-winged Teal (*Anas carolinensis**)
rb2wv2	☐	Mallard (*Anas platyrhynchos*)
pm3wv3ob	☐	Pintail (*Anas acuta*)
pm3ow	☐	Garganey (*Anas querquedula*)
as	☐	Blue-winged Teal (*Anas discors*)*
as	☐	Cape Shoveler (*Anas smithii*)*
pm4wv1ob	☐	Northern Shoveler (*Anas clypeata*)
rb3 pm3wv3	☐	Marbled Duck (*Marmaronetta angustirostris*)
rb5wv4	☐	Red-crested Pochard (*Netta rufina*)
wv2ob	☐	Pochard (*Aythya ferina*)
as	☐	Ring-necked Duck (*Aythya collaris*)*
rb5wv4	☐	Ferruginous Duck (*Aythya nyroca*)
wv3pm5	☐	Tufted Duck (*Aythya fuligula*)
as	☐	Greater Scaup (*Aythya marila*)*
wv2pm2	☐	Common Scoter (*Melanitta nigra*)
as	☐	Velvet Scoter (*Melanitta fusca*)*
as	☐	Goldeneye (*Bucephala clangula*)*
wv5	☐	Red-breasted Merganser (*Mergus serrator*)*
as	☐	Goosander (*Mergus merganser*)*
asob?	☐	Ruddy Duck (*Oxyura jamaicensis*)*
asob	☐	White-headed Duck (*Oxyura leucocephala*)*
pm2	☐	Honey-buzzard (*Pernis apivorus*)
rb3	☐	Black-shouldered Kite (*Elanus caeruleus*)
mb2pm1ow	☐	Black Kite (*Milvus migrans*)
rb5wv4	☐	Red Kite (*Milvus milvus*)
as	☐	White-tailed Eagle (*Haliaeetus albicilla*)*
rb5	☐	Lammergeier (*Gypaetus barbatus*)*
mb4pm3ow	☐	Egyptian Vulture (*Neophron percnopterus*)
as	☐	Hooded Vulture (*Necrosyrtes monachus*)*
pm4wv4ob	☐	Griffon Vulture (*Gyps fulvus*)
as	☐	Rüppell's Vulture (*Gyps rueppellii*)*
asfb	☐	Lappet-faced Vulture (*Torgos tracheliotus*)*
asfb	☐	Black Vulture (*Aegypius monachus*)*
mb3pm2ow	☐	Short-toed Eagle (*Circaetus gallicus*)
rb3wv3pm3	☐	Marsh Harrier (*Circus aeruginosus*)
wv4	☐	Hen Harrier (*Circus cyaneus*)
as	☐	Pallid Harrier (*Circus macrourus*)*
mb3pm2	☐	Montagu's Harrier (*Circus pygargus*)
rb5	☐	Dark Chanting Goshawk (*Melierax metabates*)*
rb5wv5	☐	Goshawk (*Accipiter gentilis*)
rb3wv3	☐	Sparrowhawk (*Accipiter nisus*)
wv3	☐	Common Buzzard (*Buteo buteo*)
rb2/as	☐	Long-legged Buzzard (*Buteo rufinus cirtensis / B. r. rufinus*)
as	☐	Lesser Spotted Eagle (*Aquila pomarina*)*
as	☐	Spotted Eagle (*Aquila clanga*)*
rb5	☐	Tawny Eagle (*Aquila rapax*)*
asfbob	☐	Spanish Imperial Eagle (*Aquila adalberti*)*

rb4	☐	Golden Eagle (*Aquila chrysaetos*)
mb3pm2ow	☐	Booted Eagle (*Hieraaetus pennatus*)
rb3	☐	Bonelli's Eagle (*Hieraaetus fasciatus*)
rb5pm3wv3	☐	Osprey (*Pandion haliaetus*)
mb3pm3ow	☐	Lesser Kestrel (*Falco naumanni*)
rb2wv2pm2	☐	Kestrel (*Falco tinnunculus*)
as	☐	Red-footed Falcon (*Falco vespertinus*)*
wv5	☐	Merlin (*Falco columbarius*)
mb4pm3ow	☐	Hobby (*Falco subbuteo*)
mb5	☐	Eleonora's Falcon (*Falco eleonorae*)
rb3pm?	☐	Lanner (*Falco biarmicus*)
as	☐	Saker (*Falco cherrug*)*
rb3wv4pm4	☐	Peregrine Falcon (*Falco peregrinus*)
rb3	☐	Barbary Falcon (*Falco pelegrinoides*)*
rb3	☐	Barbary Partridge (*Alectoris barbara*)
rb5	☐	Double-spurred Francolin (*Francolinus bicalcaratus*)
mb3pm3wv5	☐	Quail (*Coturnix coturnix*)
rb?4(i)	☐	Pheasant (*Phasianus colchicus*)
fb	☐	Helmeted Guineafowl (*Numida meleagris*)*
rb5	☐	Andalusian Hemipode (*Turnix sylvatica*)*
rb4wv3	☐	Water Rail (*Rallus aquaticus*)
pm3ow	☐	Spotted Crake (*Porzana porzana*)*
as	☐	Sora (*Porzana carolina*)*
pm4	☐	Little Crake (*Porzana parva*)*
mb5pm4ow	☐	Baillon's Crake (*Porzana pusilla*)*
pm5ow	☐	Corncake (*Crex crex*)*
rb2wv2	☐	Moorhen (*Gallinula chloropus*)
as	☐	Allen's Gallinule (*Porphyrula alleni*)*
rb5	☐	Purple Swamphen (*Porphyrio porphyrio*)
rb3wv1	☐	Eurasian Coot (*Fulica atra*)
rb3	☐	Red-knobbed Coot (*Fulica cristata*)
wv3	☐	Common Crane (*Grus grus*)
fb	☐	Demoiselle Crane (*Anthropoides virgo*)*
rb5wv5	☐	Little Bustard (*Tetrax tetrax*)
rb5	☐	Houbara Bustard (*Chlamydotis undulata*)
fbwv5	☐	Arabian Bustard (*Ardeotis arabs*)*
rb5wv?	☐	Great Bustard (*Otis tarda*)
as	☐	Greater Painted-snipe (*Rostratula benghalensis*)*
wv2pm2	☐	Oystercatcher (*Haematopus ostralegus*)
/rb2pm2wv2	☐	Black-winged Stilt (*Himantopus himantopus*)
pm2wv2ob	☐	Avocet (*Recurvirostra avosetta*)
rb3 pm3wv3	☐	Stone-curlew (*Burhinus oedicnemus*)
rb2bm2	☐	Cream-coloured Courser (*Cursorius cursor*)
mb4pm2ow	☐	Collared Pratincole (*Glareola pratincola*)
as	☐	Black-winged Pratincole (*Glareola nordmanni*)*
rb2pm2wv2	☐	Little Ringed Plover (*Charadrius dubius*)
pm2wv2	☐	Ringed Plover (*Charadrius hiaticula*)
as	☐	Kittlitz's Plover (*Charadrius pecuarius*)*
rb2pm2wv2	☐	Kentish Plover (*Charadrius alexandrinus*)
wv3	☐	Dotterel (*Charadrius morinellus*)

as	☐	American Golden Plover (*Pluvialis dominica*)*
wv2	☐	Golden Plover (*Pluvialis apricaria*)
pm2wv2	☐	Grey Plover (*Pluvialis squatarola*)
as	☐	Sociable Lapwing (*Chettusia gregaria*)*
as	☐	White-tailed Lapwing (*Vanellus leucurus*)*
rb5wv2	☐	Northern Lapwing (*Vanellus vanellus*)
as	☐	Great Knot (*Calidris tenuirostris*)*
pm2wv2	☐	Red Knot (*Calidris canutus*)
pm2wv2	☐	Sanderling (*Calidris alba*)
as	☐	Semipalmated Sandpiper (*Calidris pusilla*)*
pm1wv2	☐	Little Stint (*Calidris minuta*)
pm3wv4	☐	Temminck's Stint (*Calidris temminckii*)
as	☐	White-rumped Sandpiper (*Calidris fuscicollis*)*
as	☐	Pectoral Sandpiper (*Calidris melanotos*)*
pm2wv4	☐	Curlew Sandpiper (*Calidris ferruginea*)
as	☐	Purple Sandpiper (*Calidris maritima*)*
wv1pm1	☐	Dunlin (*Calidris alpina*)
as	☐	Broad-billed Sandpiper (*Limicola falcinellus*)*
as	☐	Stilt Sandpiper (*Micropalama himantopus*)*
as	☐	Buff-breasted Sandpiper (*Tryngites subruficollis*)*
pm2wv3	☐	Ruff (*Philomachus pugnax*)
wv4pm4	☐	Jack Snipe (*Lymnocryptes minimus*)
pm3wv2	☐	Common Snipe (*Gallinago gallinago*)
pm5ow	☐	Great Snipe (*Gallinago media*)*
as	☐	Long-billed Dowitcher (*Limnodromus scolopaceus*)*
wv3	☐	Woodcock (*Scolopax rusticola*)
pm1wv1	☐	Black-tailed Godwit (*Limosa limosa*)
pm2wv2	☐	Bar-tailed Godwit (*Limosa lapponica*)
pm2wv2	☐	Whimbrel (*Numenius phaeopus*)
wv5	☐	Slender-billed Curlew (*Numenius tenuirostris*)*
pm2wv2	☐	Curlew (*Numenius arquata*)
pm2wv3	☐	Spotted Redshank (*Tringa erythropus*)
pm1wv1	☐	Redshank (*Tringa totanus*)
pm4wv5	☐	Marsh Sandpiper (*Tringa stagnatilis*)
pm2wv2	☐	Greenshank (*Tringa nebularia*)
as	☐	Lesser Yellowlegs (*Tringa flavipes*)*
pm2wv3	☐	Green Sandpiper (*Tringa ochropus*)
pm2wv3	☐	Wood Sandpiper (*Tringa glareola*)
as	☐	Terek Sandpiper (*Xenus cinereus*)*
pm2wv2	☐	Common Sandpiper (*Actitis hypoleucos*)
as	☐	Spotted Sandpiper (*Actitis macularia*)*
pm2wv2	☐	Turnstone (*Arenaria interpres*)
as	☐	Wilson's Phalarope (*Steganopus tricolor*)*
as	☐	Red-necked Phalarope (*Phalaropus lobatus*)*
pm2wv2	☐	Grey Phalarope (*Phalaropus fulicarius*)
pm4wv5	☐	Pomarine Skua (*Stercorarius pomarinus*)
pm3wv3	☐	Arctic Skua (*Stercorarius parasiticus*)
as	☐	Long-tailed Skua (*Stercorarius longicaudus*)*
pm4wv4	☐	Great Skua (*Catharacta skua*)
pm3wv3	☐	Mediterranean Gull (*Larus melanocephalus*)

as	☐	Laughing Gull (*Larus atricilla*)*
as	☐	Franklin's Gull (*Larus pipixcan*)*
pm4wv4	☐	Little Gull (*Larus minutus*)
pm5ow	☐	Sabine's Gull (*Larus sabini*)*
as	☐	Bonaparte's Gull (*Larus philadelphia*)*
pm2wv2	☐	Black-headed Gull (*Larus ridibundus*)
as	☐	Grey-headed Gull (*Larus cirrocephalus*)*
rb5pm4wv4	☐	Slender-billed Gull (*Larus genei*)
b5rb5pm3wv3	☐	Audouin's Gull (*Larus audouinii*)
as	☐	Ring-billed Gull (*Larus delawarensis*)*
wv5	☐	Common Gull (*Larus canus*)
pm1wv1	☐	Lesser Black-backed Gull (*Larus fuscus*)
wv5	☐	Herring Gull (*Larus argentatus*)*
rb2	☐	Yellow-legged Gull (*Larus cachinnans*)
as	☐	Iceland Gull (*Larus glaucoides*)*
as	☐	Glaucous-winged Gull (*Larus glaucescens*)*
as	☐	Glaucous Gull (*Larus hyperboreus*)*
wv5	☐	Great Black-backed Gull (*Larus marinus*)*
wv5	☐	Kittiwake (*Rissa tridactyla*)
pm2obow	☐	Gull-billed Tern (*Sterna nilotica*)
pm3wv3	☐	Caspian Tern (*Sterna caspia*)
pm3ow	☐	Royal Tern (*Sterna maxima*)
pm3ow	☐	Lesser Crested Tern (*Sterna bengalensis*)
pm2wv2	☐	Sandwich Tern (*Sterna sandvicensis*)
pm5	☐	Roseate Tern (*Sterna dougallii*)*
pm2wv5ob	☐	Common Tern (*Sterna hirundo*)
pm5ow	☐	Arctic Tern (*Sterna paradisaea*)*
mb5	☐	Bridled Tern (*Sterna anaethetus*)*
as	☐	Sooty Tern (*Sterna fuscata*)*
mb3pm2ow	☐	Little Tern (*Sterna albifrons*)
pm3wv4ob	☐	Whiskered Tern (*Chlidonias hybridus*)
pm2ow	☐	Black Tern (*Chlidonias niger*)
pm4ow	☐	White-winged Black Tern (*Chlidonias leucopterus*)
as	☐	African Skimmer (*Rhynchops flavirostris*)*
as	☐	Guillemot (*Uria aalge*)*
wv3	☐	Razorbill (*Alca torda*)
wv4	☐	Puffin (*Fratercula arctica*)
rb5	☐	Lichtenstein's Sandgrouse (*Pterocles lichtensteinii*)
rb3	☐	Crowned Sandgrouse (*Pterocles coronatus*)
rb3mb3	☐	Spotted Sandgrouse (*Pterocles senegallus*)
rb3	☐	Black-bellied Sandgrouse (*Pterocles orientalis*)
rb3/mb3	☐	Pin-tailed Sandgrouse (*Pterocles alchata*)
rb2	☐	Rock Dove (*Columba livia*)
rb3wv4	☐	Stock Dove (*Columba oenas*)
rb2wv2	☐	Woodpigeon (*Columba palumbus*)
rb1	☐	Collared Dove (*Streptopelia decaocto*)
mb2pm1ow	☐	Turtle Dove (*Streptopelia turtur*)
rb4	☐	Laughing Dove (*Streptopelia senegalensis*)
as	☐	Namaqua Dove (*Oena capensis*)*
as/rb5(i)	☐	Ring-necked Parakeet (*Psittacula krameri*)*

pm3ob	☐	Great Spotted Cuckoo (*Clamator glandarius*)
mb3pm2	☐	Common Cuckoo (*Cuculus canorus*)
as	☐	Yellow-billed Cuckoo (*Coccyzus americanus*)*
rb2	☐	Barn Owl (*Tyto alba*)
mb3pm2ow	☐	Scops Owl (*Otus scops*)
rb3	☐	'Pharaoh' Eagle Owl (*Bubo bubo*)
rb2	☐	Little Owl (*Athene noctua*)
rb2	☐	Tawny Owl (*Strix aluco*)
rb3	☐	Long-eared Owl (*Asio otus*)
pm5wv5	☐	Short-eared Owl (*Asio flammeus*)*
rb4	☐	Marsh Owl (*Asio capensis*)
mb3pm3ow	☐	European Nightjar (*Caprimulgus europaeus*)
mb3pm3wv5	☐	Red-necked Nightjar (*Caprimulgus ruficollis*)
mb3	☐	Egyptian Nightjar (*Caprimulgus aegyptius*)
mb3pm3	☐	Alpine Swift (*Tachymarptis melba*)
wv?asob?	☐	Plain Swift (*Apus unicolor*)*
mb3pm1ow	☐	Common Swift (*Apus apus*)
mb2pm2	☐	Pallid Swift (*Apus pallidus*)
mb5pm5	☐	White-rumped Swift (*Apus caffer*)*
rb2mb2	☐	Little Swift (*Apus affinis*)
rb3wv3	☐	Kingfisher (*Alcedo atthis*)
mb4	☐	Blue-cheeked Bee-eater (*Merops persicus*)
mb2pm2ow	☐	European Bee-eater (*Merops apiaster*)
mb2pm2	☐	Roller (*Coracias garrulus*)
mb3pm3ow	☐	Hoopoe (*Upupa epops*)
pm3wv5	☐	Wryneck (*Jynx torquilla*)
rb3	☐	Levaillant's Green Woodpecker (*Picus vaillantii*)
rb2	☐	Great Spotted Woodpecker (*Dendrocopos major*)
rb5	☐	Black-crowned Sparrow-Lark (*Eremopterix nigriceps*)*
rb2	☐	Bar-tailed Desert Lark (*Ammomanes cincturus*)
rb2	☐	Desert Lark (*Ammomanes deserti*)
rb2	☐	Hoopoe Lark (*Alaemon alaudipes*)
rb3	☐	Dupont's Lark (*Chersophilus duponti*)
rb3bm?	☐	Thick-billed Lark (*Rhamphocoris clotbey*)
rb2wv?	☐	Calandra Lark (*Melanocorypha calandra*)
mb2pm1ow	☐	Short-toed Lark (*Calandrella brachydactyla*)
rb2mb?wv?	☐	Lesser Short-toed Lark (*Calandrella rufescens*)
rb2	☐	Crested Lark (*Galerida cristata*)
rb1	☐	Thekla Lark (*Galerida theklae*)
rb2wv?	☐	Woodlark (*Lullula arborea*)
rb3wv3	☐	Skylark (*Alauda arvensis*)
rb2	☐	Horned (Shore) Lark (*Eremophila alpestris*)
rb2	☐	Temminck's Horned Lark (*Eremophila bilopha*)
rb3	☐	Plain Martin (*Riparia paludicola*)
pm2ob	☐	Sand Martin (*Riparia riparia*)
rb4/mb4	☐	African Rock Martin (*Hirundo fuligula*)*
rb4wv3	☐	Crag Martin (*Hirundo rupestris*)
mb2pm1ow	☐	Barn Swallow (*Hirundo rustica*)
mb3pm3ow	☐	Red-rumped Swallow (*Hirundo daurica*)
mb3pm2ow	☐	House Martin (*Delichon urbica*)

wv5	☐	Richard's Pipit (*Anthus novaeseelandiae*)*
mb3pm3ow	☐	Tawny Pipit (*Anthus campestris*)
pm2ow	☐	Tree Pipit (*Anthus trivialis*)
wv1pm1	☐	Meadow Pipit (*Anthus pratensis*)
pm4wv4	☐	Red-throated Pipit (*Anthus cervinus*)
wv4	☐	Water Pipit (*Anthus spinoletta*)
wv5	☐	Rock Pipit (*Anthus petrosus*)*
mb2pm2wv4	☐	Yellow Wagtail (*Motacilla flava*)
as	☐	Citrine Wagtail (*Motacilla citreola*)*
rb3wv3	☐	Grey Wagtail (*Motacilla cinerea*)
rb3pm2wv1	☐	Pied/White Wagtail (*Motacilla alba*)
rb2	☐	Common Bulbul (*Pycnonotus barbatus*)
rb4	☐	Dipper (*Cinclus cinclus*)
rb3wv?	☐	Wren (*Troglodytes troglodytes*)
wv5	☐	Dunnock (*Prunella modularis*)*
rb4wv4	☐	Alpine Accentor (*Prunella collaris*)
mb3pm3	☐	Rufous Bush Robin (*Cercotrichas galactotes*)
rb3wv1	☐	Robin (*Erithacus rubecula*)
mb2pm2	☐	Nightingale (*Luscinia megarhynchos*)
pm3wv4	☐	Bluethroat (*Luscinia svecica*)
rb4wv2	☐	Black Redstart (*Phoenicurus ochruros*)
mb4pm2ow	☐	Redstart (*Phoenicurus phoenicurus*)
rb3	☐	Moussier's Redstart (*Phoenicurus moussieri*)
pm3ow	☐	Whinchat (*Saxicola rubetra*)
rb3wv1	☐	Stonechat (*Saxicola torquata*)
pm5	☐	Isabelline Wheatear (*Oenanthe isabellina*)*
mb3pm2ow	☐	Northern Wheatear (*Oenanthe oenanthe*)
mb2pm2	☐	Black-eared Wheatear (*Oenanthe hispanica*)
rb2mb2	☐	Desert Wheatear (*Oenanthe deserti*)
rb3mb?	☐	Red-rumped Wheatear (*Oenanthe moesta*)
rb4	☐	Mourning Wheatear (*Oenanthe lugens*)
rb2	☐	White-crowned Black Wheatear (*Oenanthe leucopyga*)
rb2	☐	Black Wheatear (*Oenanthe leucura*)
mb4pm4ow	☐	Rock Thrush (*Monticola saxatilis*)
rb3wv3	☐	Blue Rock Thrush (*Monticola solitarius*)
wv2	☐	Ring Ouzel (*Turdus torquatus*)
rb1wv?	☐	Blackbird (*Turdus merula*)
as	☐	Fieldfare (*Turdus pilaris*)*
wv2	☐	Song Thrush (*Turdus philomelos*)
wv3	☐	Redwing (*Turdus iliacus*)
rb3wv4	☐	Mistle Thrush (*Turdus viscivorus*)
rb2wv?	☐	Cetti's Warbler (*Cettia cetti*)
rb2wv4	☐	Zitting Cisticola (Fan-tailed Warbler) (*Cisticola juncidis*)
rb3	☐	Scrub Warbler (*Scotocerca inquieta*)
pm4wv5	☐	Grasshopper Warbler (*Locustella naevia*)
as	☐	River Warbler (*Locustella fluviatilis*)*
mb4pm4	☐	Savi's Warbler (*Locustella luscinioides*)
rb4pm5wv5	☐	Moustached Warbler (*Acrocephalus melanopogon*)
pm5	☐	Aquatic Warbler (*Acrocephalus paludicola*)*
pm2ow?	☐	Sedge Warbler (*Acrocephalus schoenobaenus*)

mb3pm1ow	☐	Reed Warbler (*Acrocephalus scirpaceus*)
mb3pm4	☐	Great Reed Warbler (*Acrocephalus arundinaceus*)
mb2pm3ow	☐	Western Olivaceous Warbler (*Hippolais opaca*)
mb5	☐	Eastern Olivaceous Warbler (*Hippolais pallida*)
as/pm?	☐	Icterine Warbler (*Hippolais icterina*)*
mb3pm3ow	☐	Melodious Warbler (*Hippolais polyglotta*)
as	☐	Marmora's Warbler (*Sylvia sarda*)*
rb3wv3	☐	Dartford Warbler (*Sylvia undata*)
rb3mb3	☐	Tristram's Warbler (*Sylvia deserticola*)
rb/mb3pm3	☐	Spectacled Warbler (*Sylvia conspicillata*)
mb3pm2ow	☐	Subalpine Warbler (*Sylvia cantillans*)
rb1	☐	Sardinian Warbler (*Sylvia melanocephala*)
rb/mb4	☐	Desert Warbler (*Sylvia nana*)
mb4pm3ow	☐	Orphean Warbler (*Sylvia hortensis*)
pm5ow	☐	Lesser Whitethroat (*Sylvia curruca*)*
mb4pm3ow	☐	Whitethroat (*Sylvia communis*)
pm2ow	☐	Garden Warbler (*Sylvia borin*)
rb4pm1wv1	☐	Blackcap (*Sylvia atricapilla*)
as	☐	Pallas's Warbler (*Phylloscopus proregulus*)*
as	☐	Yellow-browed Warbler (*Phylloscopus inornatus*)*
as	☐	Dusky Warbler (*Phylloscopus fuscatus*)*
mb3pm3ow	☐	Western Bonelli's Warbler (*Phylloscopus bonelli*)
pm4	☐	Wood Warbler (*Phylloscopus sibilatrix*)
wv1pm2	☐	Common Chiffchaff (*Phylloscopus collybita*)
rb5pm3wv?	☐	Iberian Chiffchaff (*Phylloscopus ibericus*)*
pm1ow	☐	Willow Warbler (*Phylloscopus trochilus*)
as	☐	Goldcrest (*Regulus regulus*)*
rb3wv4	☐	Firecrest (*Regulus ignicapillus*)
mb3pm3ow	☐	Spotted Flycatcher (*Muscicapa striata*)
as	☐	Red-breasted Flycatcher (*Ficedula parva*)*
as	☐	Collared Flycatcher (*Ficedula albicollis*)*
pm2ow	☐	Pied Flycatcher (*Ficedula hypoleuca*)
mb4	☐	Atlas Flycatcher (*Ficedula speculigera*)
as	☐	Bearded Tit (*Panurus biarmicus*)*
rb3	☐	Fulvous Babbler (*Turdoides fulvus*)
as	☐	Long-tailed Tit (*Aegithalos caudatus*)*
as	☐	Crested Tit (*Parus cristatus*)*
rb2ow	☐	Coal Tit (*Parus ater*)
rb1ow?	☐	Blue Tit (*Parus caeruleus*)
rb2ow	☐	Great Tit (*Parus major*)
rb3	☐	Nuthatch (*Sitta europaea*)
as	☐	Wallcreeper (*Tichodroma muraria*)*
rb3	☐	Short-toed Treecreeper (*Certhia brachydactyla*)
as	☐	Penduline Tit (*Remiz pendulinus*)*
mb3pm3	☐	Golden Oriole (*Oriolus oriolus*)
rb3	☐	Black-crowned Tchagra (*Tchagra senegala*)
as	☐	Red-backed Shrike (*Lanius collurio*)*
rb2wv?	☐	Southern Grey Shrike (*Lanius meridionalis*)
mb2pm2	☐	Woodchat Shrike (*Lanius senator*)
rb3	☐	Jay (*Garrulus glandarius*)

rb3 ☐	Magpie (*Pica pica*)
rb3wv? ☐	Alpine Chough (*Pyrrhocorax graculus*)
rb3 ☐	Red-billed Chough (*Pyrrhocorax pyrrhocorax*)
rb3 ☐	Jackdaw (*Corvus monedula*)
ow ☐	Carrion Crow (*Corvus corone*)*
rb3 ☐	Brown-necked Raven (*Corvus ruficollis*)
rb3 ☐	Common Raven (*Corvus corax*)
wv2 ☐	Starling (*Sturnus vulgaris*)
rb3wv5 ☐	Spotless Starling (*Sturnus unicolor*)
rb1wv ☐	House Sparrow (*Passer domesticus*)
rb2pmwv ☐	Spanish Sparrow (*Passer hispaniolensis*)
rb5 ☐	Desert Sparrow (*Passer simplex*)
wv5ob ☐	Tree Sparrow (*Passer montanus*)
rb2 ☐	Rock Sparrow (*Petronia petronia*)
as ☐	Red-billed Firefinch (*Lagonosticta senegala*)*
as ☐	Red-eyed Vireo (*Vireo olivaceus*)*
rb2wv2 ☐	Chaffinch (*Fringilla coelebs*)
wv5 ☐	Brambling (*Fringilla montifringilla*)
rb2wv2 ☐	Serin (*Serinus serinus*)
as ☐	Citril Finch (*Serinus citrinella*)*
rb2wv2 ☐	Greenfinch (*Carduelis chloris*)
rb1wv1 ☐	Goldfinch (*Carduelis carduelis*)
wv4 ☐	Siskin (*Carduelis spinus*)
rb2wv2 ☐	Linnet (*Carduelis cannabina*)
as ☐	Redpoll (*Carduelis flammea*)*
rb3wv5 ☐	Crossbill (*Loxia curvirostra*)
rb4 ☐	Crimson-winged Finch (*Rhodopechys sanguinea*)
rb2 ☐	Trumpeter Finch (*Bucanetes githagineus*)
as ☐	Scarlet Rosefinch (*Carpodacus erythrinus*)*
as ☐	Bullfinch (*Pyrrhula pyrrhula*)*
rb3wv4 ☐	Hawfinch (*Coccothraustes coccothraustes*)
as ☐	Snow Bunting (*Plectophenax nivalis*)*
as ☐	Yellowhammer (*Emberiza citrinella*)*
rb3wv? ☐	Cirl Bunting (*Emberiza cirlus*)
rb3wv4 ☐	Rock Bunting (*Emberiza cia*)
rb2 ☐	House Bunting (*Emberiza striolata*)
pm4 ☐	Ortolan Bunting (*Emberiza hortulana*)
rb5wv4 ☐	Reed Bunting (*Emberiza schoeniclus*)
as ☐	Black-headed Bunting (*Emberiza melanocephala*)*
rb2wv2 ☐	Corn Bunting (*Miliaria calandra*)

ODONATES

after Jacquemin & Boudot 1999

ZYGOPTERES

Calopterygidae Buchecker, 1876

- ☐ *Calopteryx exul* Sélys, 1853
- ☐ *Calopteryx h. haemorrhoidalis* (Van der Linden, 1825)
- ☐ *Calopteryx virgo* (Linne, 1758) ssp. *meridionalis* Sélys, 1873

Lestidae Sélys, 1840

- ☐ *Lestes barbarus* (Fabricius, 1798)
- ☐ *Lestes v. virens* (Charpentier, 1825)
- ☐ *Lestes v. viridis* Van der Linden, 1825
- ☐ *Lestes dryas* Kirby, 1890
- ☐ *Lestes macrostigma* (Eversmann, 1836)
- ☐ *Sympecma fusca* (Van der Linden, 1820)

Platycnemididae Tillyard, 1938

- ☐ *Platycnemis subdi latata* Selys, 1849

Coenagrionidae Kirby, 1890

- ☐ *Coenagrion puella*(Linné, 1758)
- ☐ *Coenagrion m. mercuriale* (Charpentier, 1840)
- ☐ *Coenagrion caerulescens* (Fonscolombe, 1838)
- ☐ *Coenagrion scitulum* (Sélys, 1842)
- ☐ *Cercion l. lindenii* (Sélys, 1840)
- ☐ *Enallagma cyathigerum* (Charpentier, 1840)
- ☐ *Enallagma deserti* (Sélys, 1871)
- ☐ *Ischnura pumilio* (Charpentier, 1825)
- ☐ *Ischnura graellsii* (Rambur, 1842)
- ☐ *Ischnura saharensis* Aguesse, 1958
- ☐ *Ischnura fountaineae* Morton, 1905
- ☐ *Erythromma viridulum* (Charpentier, 1840)
- ☐ *Pyrrhosoma n. nymphula* (Sulzer, 1776)
- ☐ *Ceriagrion t. tenellum* (Villers, 1789)
- ☐ *Pseudagrion s. sublacteum* (Karsch, 1893)

ANISOPTERES

Aeshnidae Sélys, 1850

- ☐ *Aeshna affiinis* Van der Linden, 1823
- ☐ *Aeshna mixta* Latreille, 1805
- ☐ *Aeshna cyanea* (Müller, 1764)
- ☐ *Aeshna i. isoceles* (Müller, 1767)
- ☐ *Boyeria irene* (Fonscolombe, 1838)
- ☐ *Hemianax ephippiger* (Burmeister, 1839)
- ☐ *Anax imperator* Leach, 1815
- ☐ *Anax parthenope* Sélys, 1839

Gomphidae Sélys, 1850

- ☐ *Gomphus simillimus* Sélys, 1840 ssp. *maroccanus* Lieftinck, 1966

- ☐ *Onychogompnus forcipatus* (Linné, 1758) ssp. *unguiculatus* (Vander Linden, 1823)
- ☐ *Onychogompnus uncatus* (Charpentier, 1840)
- ☐ *Onychogomphus costae* Sélys, 1885
- ☐ *Paragomphus genei* (Sélys, 1841)

Cordulegastridae Fraser, 1940

- ☐ *Cordulegaster boltonii* (Donovan, 1807) ssp. *algirica* Morton, 1915
- ☐ *Cordulegaster princeps* Morton, 1915

Corduliidae Tillyard, 1926

- ☐ *Oxygastra curtisii* (Dale, 1834)

Libellulidae Sélys, 1850

- ☐ *Libellula q. quadrimaculata* Linné, 1758
- ☐ *Orthetrum c. cancellatum* (Linné, 1758)
- ☐ *Orthetrum brunneum* (Fonscolombe, 1837)
- ☐ *Orthetrum coerulescens* (Fabricius, 1798) ssp. *anceps* (Schneider, 1845)
- ☐ *Orthetrum nitidinerve* (Sélys, 1841)
- ☐ *Orthetrum c. chrysostigma* (Burmeister, 1832)
- ☐ *Orthetrum t. trinacria* (Sélys, 1841)
- ☐ *Brachythemis leucosticta* (Burmeister, 1839)
- ☐ *Crocothemis e. erythraea* (Brulle, 1832)
- ☐ *Diplacodes lefebvrii* (Rambur, 1842)
- ☐ *Sympetrum s. sanguineum* (Müller, 1764)
- ☐ *Sympetrum s. striolatum* (Charpentier, 1840)
- ☐ *Sympetrum meridionale* (Sélys, 1841)
- ☐ *Sympetrum fonscolombii* (Sélys , 1840)
- ☐ *Trithemis annulata* (Palisot de Beauvois, 1805)
- ☐ *Trithemis a. arteriosa* (Burmeister, 1839)
- ☐ *Trithemis kirbyi* Sélys, 1891
- ☐ *Zygonyx t. torridus* (Kirby, 1889)
- ☐ *Pantala flavescens*(Fabricius, 1798)

MAMMALS

adapted from Aulagnier and Thévenot 1986

This list excludes the species that have disappeared from Morocco or whose occurrence is not currently documented well enough.

Insectivora

- ☐ Algerian Hedgehog (*Erinaceus algirus*)
- ☐ Desert Hedgehog (*Paraechinus aethiopicus*)
- ☐ North African Lesser White-toothed Shrew (*Crocidura whitakeri*)
- ☐ Tarfaya's Shrew (*Crocidura tarfayaensis*)
- ☐ Greater White-toothed Shrew (*Crocidura russula*)
- ☐ (*Crocidura lusitania*)
- ☐ (*Crocidura bolivari*)
- ☐ Pigmy White-toothed Shrew (*Suncus etruscus*)

Macroscelidea

- ☐ North African Elephant Shrew (*Elephantulus rozeti*)

Chiroptera

- ☐ Larger Rat-tailed Bat (*Rhinopoma microphyllum*)
- ☐ Lesser Rat-tailed Bat (*Rhinopoma hardwickei*)
- ☐ Egyptian Slit-faced Bat (*Nycteris thebaica*)
- ☐ Greater Horseshoe Bat (*Rhinolophus ferrumequinum*)
- ☐ Lesser Horseshoe Bat (*Rhinolophus hipposideros*)
- ☐ Mediterranean Horseshoe Bat (*Rhinolophus euryale*)
- ☐ Mehely's Horseshoe Bat (*Rhinolophus mehely*)
- ☐ Blasius's Horseshoe Bat (*Rhinolophus blasii*)
- ☐ Sundevall's African Leaf-nosed Bat (*Hipposideros caffer*)
- ☐ Trident Leaf-nosed Bat (*Asellia tridens*)
- ☐ Whiskered Bat (*Myotis mystacinus*)
- ☐ Geoffroy's Bat (*Myotis emarginatus*)
- ☐ Natterer's Bat (*Myotis nattereri*)
- ☐ Long-fingered Bat (*Myotis capaccinii*)
- ☐ Lesser Mouse-eared Bat (*Myotis blythi*)
- ☐ Common Pipistrelle (*Pipistrellus pipistrellus*)
- ☐ Kuhl's Pipistrelle (*Pipistrellus kuhli*)
- ☐ Savi's Pipistrelle (*Pipistrellus savii*)
- ☐ Ruppell's Bat (*Pipistrellus rueppelli*)
- ☐ Greater Noctule (*Nyctalus lasiopterus*)
- ☐ Serotine (*Eptesicus serotinus*)
- ☐ Hemprich's Long-eared Bat (*Otonycteris hemprichi*)
- ☐ Barbastelle (*Barbastella barbastellus*)
- ☐ Grey Long-eared Bat (*Plecotus austriacus*)
- ☐ Schreiber's Bat (*Miniopterus schreibersi*)
- ☐ Free-tailed Bat (*Tadarida teniotis*)

Primates

- ☐ Barbary Ape (*Macaca sylvanus*)

Lagomorpha

- ☐ Brown Hare (*Lepus capensis*)
- ☐ Rabbit (*Oryctolagus cuniculus*)

Rodentia

- ☐ Barbary Ground-squirrel (*Atlantoxerus getulus*)
- ☐ Geoffroy's Ground-squirrel (*Xerus erythropus*)
- ☐ Large North African Gerbil (*Gerbillus campestris*)
- ☐ Baluchistan Gerbil (*Gerbillus nanus*)
- ☐ Pigmy Gerbil (*Gerbillus henleyi*)
- ☐ Lesser Egyptian Gerbil (*Gerbillus gerbillus*)
- ☐ Greater Egyptian Gerbil (*Gerbillus pyramidum*)
- ☐ (*Gerbillus hesperinus*)
- ☐ (*Gerbillus hoogstraali*)
- ☐ (*Gerbillus occiduus*)
- ☐ (*Gerbillus riggenbachi*)
- ☐ Lesser Short-tailed Gerbil (*Dipodillus simoni*)
- ☐ Greater Short-tailed Gerbil (*Dipodillus maghrebi*)
- ☐ Fat-tailed Gerbil (*Pachyuromys duprasi*)
- ☐ Shaw's Jird (*Meriones shawi*)
- ☐ Lybian Jird (*Meriones libycus*)
- ☐ Sundevall's Jird (*Meriones crassus*)
- ☐ Fat Sand Rat (*Psammomys obesus*)
- ☐ Wood Mouse (*Apodemus sylvaticus*)
- ☐ Barbary Striped Mouse (*Lemniscomys barbarus*)
- ☐ Black Rat (*Rattus rattus*)
- ☐ Brown Rat (*Rattus norvegicus*)
- ☐ House Mouse (*Mus musculus*)
- ☐ Algerian Mouse (*Mus spretus*)
- ☐ Western Multimammate Rat (*Mastomys erythroleucus*)
- ☐ Spiny Mouse (*Acomys cahirinus*)
- ☐ Garden Dormouse (*Elyomis quercinus*)
- ☐ Lesser Egyptian Jerboa (*Jaculus jaculus*)
- ☐ Greater Egyptian Jerboa (*Jaculus orientalis*)
- ☐ Porcupine (*Hystrix cristata*)
- ☐ Gundi (*Ctenodactylus gundi*)
- ☐ Thomas's Gundi (*Ctenodactylus vali*)

Carnivora

- ☐ Jackal (*Canis aureus*)
- ☐ Common Red Fox (*Vulpes vulpes*)
- ☐ Sand Fox (*Vulpes rueppelli*)
- ☐ Fennec (*Fennecus zerda*)
- ☐ Weasel (*Mustela nivalis*)
- ☐ Ferret (*Mustela putorius furo*)
- ☐ Saharan Striped-Weasel (*Poecilictis libyca*)
- ☐ Ratel (*Mellivora capensis*)
- ☐ Otter (*Lutra lutra*)
- ☐ Genet (*Genetta genetta*)
- ☐ Egyptian Mongoose (*Herpestes ichneumon*)
- ☐ Striped Hyaena (*Hyaena hyaena*)
- ☐ African Wild Cat (*Felis libyca*)
- ☐ Sand Cat (*Felis margarita*)
- ☐ Caracal (*Felis caracal*)

- ☐ Leopard (*Panthera pardus*)
- ☐ Cheetah (*Acinonyx jubatus*)

Artiodactyla

- ☐ Wild Boar (*Sus scrofa*)
- ☐ Dorcas Gazelle (*Gazella dorcas*)
- ☐ Edmi Gazelle (*Gazella cuvieri*)
- ☐ Addra Gazelle (*Gazella dama*)
- ☐ Barbary Sheep (*Ammotragus lervia*)
- ☐ Red Deer (*Cervus elaphus*)

SEA MAMMALS

after Bayed and Beaubrun 1987

Cetacea

- ☐ Sei Whale (*Balaenoptera borealis*)
- ☐ Fin Whale (*Balaenoptera physalus*)
- ☐ Blue Whale (*Balaenoptera musculus*)
- ☐ Humpback (*Megaptera novaeangliae*)
- ☐ Striped Dolphin (*Stenella coeruleoalba*)
- ☐ Common Dolphin (*Delphinus delphis*)
- ☐ (*Sousa teuszii*)
- ☐ Bottlenose Dolphin (*Tursiops truncatus*)
- ☐ False Killer Whale (*Pseudorca crassidens*)
- ☐ Killer Whale (*Orcinus orca*)
- ☐ Risso's Dolphin (*Grampus griseus*)
- ☐ Long-finned Pilot Whale (*Globicephala melaena*)
- ☐ Harbour Porpoise (*Phocoena phocoena*)
- ☐ Sperm Whale (*Physeter macrocephalus*)
- ☐ Cuvier's Beaked Whale (*Ziphius cavirostris*)
- ☐ Northern Bottlenose Whale (*Hyperoodon ampullatus*)

Pinnipedia

- ☐ Mediterranean Monk Seal (*Monachus monachus*)

AMPHIBIANS AND REPTILES

after Bons and Geniez 1996, and Geniez, Mateo and Bons 2000

Urodela

☐ Sharp-ribbed Newt (*Pleurodeles waltl*)
☐ Fire Salamander (*Salamandra salamandra*)

Anoura

☐ Painted Frog (*Discoglossus pictus*)
☐ Midwife Toad (*Alytes obstetricans*)
☐ Moroccan Spadefoot (*Pelobates varaldii*)
☐ Common Toad (*Bufo bufo*)
☐ Mauritanian Toad (*Bufo mauritanicus*)
☐ Green Toad (*Bufo viridis*)
☐ (*Bufo xeros*)
☐ Brongersma's Toad (*Bufo brongersmai*)
☐ Stripeless Tree Frog (*Hyla meridionalis*)
☐ North African Green Frog (*Rana saharica*)
☐ (*Dicroglossus occipitalis*)

Chelonia

☐ Spur-thighed Tortoise (*Testudo graeca*)
☐ European Pond Terrapin (*Emys orbicularis*)
☐ Spanish Terrapin (*Mauremys leprosa*)
☐ Common Green Turtle (*Chelonia mydas*)
☐ Loggerhead Turtle (*Caretta caretta*)
☐ Olive Ridley Turtle (*Lepidochelys olivacea*)
☐ Hawksbill Turtle (*Eretmochelys imbricata*)
☐ Leathery Turtle (*Dermochelys coriacea*)

Sauria

☐ Moorish Gecko (*Tarentola mauritanica*)
☐ Böhme's Gecko (*Tarentola boehmei*)
☐ Desert Gecko (*Tarentola deserti*)
☐ House Gecko (*Tarentola annularis*)
☐ Hoggar Gecko (*Tarentola ephippiata*)
☐ Turkish Gecko (*Hemidactylus turcicus*)
☐ Oudri's Fan-footed Gecko (*Ptyodactylus oudrii*)
☐ Helmeted Gecko (*Geckonia chazaliae*)
☐ High Atlas Moroccan Day Gecko (*Quedenfeldtia trachyblepharus*)
☐ Moroccan Day Gecko (*Quedenfeldtia moerens*)
☐ Elegant Gecko (*Stenodactylus sthenodactylus*)
☐ Petrie's Gecko (*Stenodactylus petrii*)
☐ Toed Gecko (*Saurodactylus mauritanicus*)
☐ Brosset's Toed Gecko (*Saurodactylus brosseti*)
☐ Banded Toed Gecko (*Saurodactylus fasciatus*)
☐ Tripoli Pigmy Gecko (*Tropiocolotes tripolitanus*)
☐ Bibron's Agama (*Agama bibroni*)
☐ Changeable Agama (*Trapelus mutabilis*)
☐ (*Uromastix flavifasciata*)
☐ (*Uromastix occidentalis*)
☐ Spiny-tailed Agama (*Uromastyx acanthinurus*)

- ☐ Common Chameleon (*Chamaeleo chamaeleon*)
- ☐ Moroccan Glass Lizard (*Ophisaurus koellikeri*)
- ☐ Desert Monitor (*Varanus griseus*)
- ☐ North African Eyed Lizard (*Lacerta pater*)
- ☐ Andreanszky's Lizard (*Lacerta andreanszkyi*)
- ☐ Moroccan Rock Lizard (*Scelarcis perspicillata*)
- ☐ Iberian Wall Lizard (*Podarcis hispanica*)
- ☐ Large Psammodromus (*Psammodromus algirus*)
- ☐ Blanc's Psammodromus (*Psammodromus blanci*)
- ☐ Green Psammodromus (*Psammodromus microdactylus*)
- ☐ Spanish Psammodromus (*Psammodromus hispanus*)
- ☐ Mograbin Snake-eyed Lizard (*Ophisops occidentalis*)
- ☐ Small Spotted Lizard (*Mesalina guttulata*)
- ☐ Olivier's Small Lizard (*Mesalina olivieri*)
- ☐ Pasteur's Small Lizard (*Mesalina pasteuri*)
- ☐ Red-spotted Small Lizard (*Mesalina rubropunctata*)
- ☐ Common Fringe-toed Lizard (*Acanthodactylus erythrurus*)
- ☐ Coastal Common Fringe-toed Lizard (*Acanthodactylus lineomaculatus*)
- ☐ Savigny's Fringe-toed Lizard (*Acanthodactylus savignyi*)
- ☐ Spotted Leopard Fringe-toed Lizard (*Acanthodactylus maculatus*)
- ☐ Busack's Leopard Fringe-toed Lizard (*Acanthodactylus busacki*)
- ☐ Bosc's Fringe-toed Lizard (*Acanthodactylus boskianus*)
- ☐ Dumeril's Fringe-toed Lizard (*Acanthodactylus dumerili*)
- ☐ Long-footed Fringe-toed Lizard (*Acanthodactylus longipes*)
- ☐ Golden Fringe-toed Lizard (*Acanthodactylus aureus*)
- ☐ Algerian Orange-tailed Skink (*Eumeces algeriensis*)
- ☐ Ocellated Skink (*Chalcides ocellatus*)
- ☐ Manuel's Skink (*Chalcides manueli*)
- ☐ Riffian Skink (*Chalcides colosii*)
- ☐ Ebner's Skink (*Chalcides ebneri*)
- ☐ Mountain Skink (*Chalcides montanus*)
- ☐ Doumergue's Skink (*Chalcides parallelus*)
- ☐ Many-scaled Skink (*Chalcides polylepis*)
- ☐ Mionecton Skink (*Chalcides mionecton*)
- ☐ Moroccan Three-toed Skink (*Chalcides pseudostriatus*)
- ☐ Small Three-toed Skink (*Chalcides minutus*)
- ☐ Two-fingered Skink (*Chalcides mauritanicus*)
- ☐ De l'Isles's Sand Skink (*Sphenops delislii*)
- ☐ Boulenger's Sand Skink (*Sphenops boulengeri*)
- ☐ Senegal Sand Skink (*Sphenops sphenopsiformis*)
- ☐ White-banded Sandfish (*Scincus albifasciatus*)
- ☐ Banded Skink (*Scincopus fasciatus*)

Amphisbaenia

- ☐ Tangier Worm Lizard (*Blanus tingitanus*)
- ☐ Moroccan Worm Lizard (*Blanus mettetali*)
- ☐ Sharp-tailed Worm Lizard (*Trogonophis wiegmanni*)

Serpentes

- ☐ Beaked Thread Snake (*Leptotyphlops macrorhynchus*)
- ☐ Javeline Sand Boa (*Eryx jaculus*)

- ☐ Horseshoe Snake (*Coluber hippocrepis*)
- ☐ Algerian Snake (*Coluber algirus*)
- ☐ Diadem Snake (*Spalerosophis diadema*)
- ☐ Mograbin Diadem Snake (*Spalerosophis dolichospilus*)
- ☐ Southem Smooth Snake (*Coronella girondica*)
- ☐ Hooded Snake (*Macroprotodon cucullatus*)
- ☐ North African Catsnake (*Telescopus dhara*)
- ☐ Common Leaf-nosed Snake (*Lytorhynchus diadema*)
- ☐ Common African House Snake (*Lamprophis fuliginosus*)
- ☐ Egg-eating Snake (*Dasypeltis scabra*)
- ☐ Viperine Snake (*Natrix maura*)
- ☐ Grass Snake (*Natrix natrix*)
- ☐ Montpellier Snake (*Malpolon monspessulanus*)
- ☐ Moila Snake (*Malpolon moilensis*)
- ☐ Schokar Sand Snake (*Psammophis schokari*)
- ☐ Egyptian Cobra (*Naja haje*)
- ☐ Lataste's Viper (*Vipera latastei*)
- ☐ Mountain Viper (*Vipera monticola*)
- ☐ Moorish Viper (*Macrovipera mauritanica*)
- ☐ Greater Cerastes (*Cerastes cerastes*)
- ☐ Lesser Cerastes (*Cerastes vipera*)
- ☐ Puff Adder (*Bitis arietans*)
- ☐ White-bellied Carpet Viper (*Echis leucogaster*)

ORCHIDS

after Raynaud 1985

- ☐ Yellow Bee Ophrys (*Ophrys lutea*)
- ☐ Sombre Bee Ophrys (*Ophrys fusca*)
- ☐ Atlantic Bee Ophrys (*Ophrys atlantica*)
- ☐ Moroccan Bee Ophrys (*Ophrys dyris*)
- ☐ Mirror Ophrys (*Ophrys vernixia*)
- ☐ Sawfly Ophrys (*Ophrys tenthredinifera*)
- ☐ Moroccan Woodcock Ophrys (*Ophrys scolopax apiformis*)
- ☐ Bee Ophrys (*Ophrys apifera*)
- ☐ Bumble Bee Ophrys (*Ophrys bombyliflora*)
- ☐ Marsh Lax-flowered Orchid (*Orchis palustris*)
- ☐ Early Purple Orchid (*Orchis mascula s.l.*)
- ☐ Fan-lipped Orchid (*Orchis saccata*)
- ☐ Pink Butterfly Orchid (*Orchis papilionacea*)
- ☐ Spitzel's Orchid (*Orchis spitzelii*)
- ☐ Green-winged Orchid (*Orchis champagneuxii*)
- ☐ Bug Orchid (*Orchis coriophora*)
- ☐ Milky Orchid (*Orchis lactea*)
- ☐ Naked Man Orchid (*Orchis italica*)
- ☐ Markusi's Marsh Orchid (*Dactylorhiza markusii*)
- ☐ Moorish Marsh Orchid (*Dactylorhiza maurusia*)
- ☐ Robust Marsh Orchid (*Dactylorhira elata elata*)
- ☐ Robust Marsh Orchid (*Dactylorhiza elata durandii*)
- ☐ Tongue Orchid (*Serapias lingua*)
- ☐ Small-flowered Tongue Orchid (*Serapias parviflora*)
- ☐ Long-lipped Tongue Orchid (*Serapias vomeracea*)
- ☐ Heart-flowered Tongue Orchid (*Serapias cordigera*)
- ☐ Man Orchid (*Aceras anthropophorum*)
- ☐ Giant Orchid (*Barlia robertiana*)
- ☐ Lizard Orchid (*Himantoglossum hircinum*)
- ☐ Pyramidal Orchid (*Anacamptis pyramidalis*)
- ☐ Greater Butterfly Orchid (*Platanthera chlorantha*)
- ☐ Algerian Butterfly Orchid (*Platanthera algeriensis*)
- ☐ Two-leaved Scrub Orchid (*Gennaria diphylla*)
- ☐ Dense-flowered Orchid (*Neotinea maculata*)
- ☐ Broad-leaved Helleborine (*Epipactis helleborine*)
- ☐ Red Helleborine (*Cephalantera rubra*)
- ☐ Sword-leaved Helleborine (*Cephalantera longifolia*)
- ☐ Violet Limodore (*Limodorum abortivum*)
- ☐ Autumn Lady's Tresses (*Spiranthes spiralis*)
- ☐ Summer Lady's Tresses (*Spiranthes aestivalis*)

SELECTED BIBLIOGRAPHY

Birds – Field Guides

No specific Moroccan field guide was available in 2002. Those most useful in Morocco are:

The Collins Bird Guide by **Svensson, L., Mullarney, K., Zetterstrom, D. & Grant, P.J.** HarperCollins.

Birds of Europe with North Africa and the Middle East by **Jonsson, L.** Christopher Helm.

Birds of Britain and Europe with the Middle East and North Africa by **Heinzel, H., Fitter, R. and Parslow, J.** Collins.

The Handbook of Bird Identification by **Beaman, M. & Madge, S.** Christopher Helm.

Birds – General

As part of the Western Palearctic, Morocco is included in the series *Handbook of the Birds of Europe, the Middle East and North Africa – The Birds of the Palearctic* published between 1977 and 1994 and in the subsequent *The Birds of the Palearctic – Concise Edition* published in 1998 both by Oxford University Press.

Morocco is also covered in the multi-volume series *The Birds of Africa* the first volume of which appeared in 1982 (Academic Press).

Birds – Regional

Etchécopar, R.D. & Hüe, F. (translated by P.A.D. Hollom). 1967. *The Birds of North Africa*. Oliver & Boyd, Edinburgh & London.

Finlayson, J.C. 1992. *Birds of the Straits of Gibraltar*. T. & A.D. Poyser. London, UK. 534 pp.

Heim de Balsac, H. & Mayaud, N. 1962. *Les Oiseaux du Nord-Ouest de l'Afrique. Distribution géographique, Ecologie, Migrations, Reproduction.* Encyclopédie ornithologique X, Lechevalier, Paris. 487 pp.

Isenmann, P. & Moali, A. 2000. *Oiseaux d'Algérie/Birds of Algeria*. Soc. Etudes Ornith. Fr. Paris, France. 336 pp. ISBN 2-9506548-8-6.

Birds – Morocco

Thévenot, M., Vernon, J.D.R. & Bergier, P. 2003. *The Birds of Morocco.* BOU Checklist series.

Birds – Cross-references

This section only lists those refences cited in the text of this Guide and some key papers related to the sites described in this Guide. For a comprehensive list of ornithological publications, see Thévenot, Vernon and Bergier (2003).

Barreau, D., Bergier, P. & Lesne, L. 1987. L'avifaune de l'Oukaimeden, 2,200-3,600m (Haut Atlas, Maroc). *L'Oiseau et la R.F.O.* 57: 307-367.

Barreau, D. & Bergier, P. 2000-2001. L'avifaune de la région de Marrakech (Haouz et Haut Atlas de Marrakech, Maroc). 1. Le cadre. *Alauda* 68: 301-310; 2. Les espèces: non passereaux. *Alauda* 69: 167-202; 3. Les espèces: passereaux. *Alauda* 69: 261-309.

Bergier, P. 1987. *Les Rapaces diurnes du Maroc. Statut, répartition et écologie.* Annales du Centre d'Etude sur les Ecosystèmes de Provence (C.E.E.P.) Aix en Provence 3: 160 pp.

Bergier, P., Franchimont, J. & Thévenot, M. 1999. Implantation et expansion géographique de deux espèces de Columbidés au Maroc : la Tourterelle turque *Streptopelia decaocto* et la Tourterelle maillée *Streptopelia senegalensis*. *Alauda* 67: 23-36.

Bergier, P & Thévenot, M. 1991. Statut et écologie du Hibou du Cap nord-africain *Asio capensis tingitanus*. *Alauda* 59: 206-224.

El Hamoumi, R., Dakki, M. & Thévenot, M. 2000. Composition et phénologie du peuplement d'Oiseaux d'eau du complexe lagunaire de Sidi Moussa-Walidia (Maroc). Son importance nationale et internationale. *Alauda* 68: 275-294.

Pineau, J. & Giraud-Audine, M. 1979. Les oiseaux de la Péninsule Tingitane. Bilan des connaissances actuelles. *Trav. Inst. Sci. Rabat, ser. Zool.* 38: 1-147.

Schollaert, V. & Willem, G. 2000. Taxonomy of the Peregrine *Falco peregrinus* / Barbary Falcon *F. (peregrinus) pelegrinoides* complex in Morocco. *African Bird Club Bull.* 7: 101-103.

Porphyrio (the journal of the GOMAC) publishes the annual ornithological reports since 1989 and the Moroccan Rare Birds Committee reports since 1995.

Birds – Sound Guides

Oiseaux d'Afrique (I). Sahara, Maghreb, Madère, Canaries et Iles du Cap Vert 2000. Société d'Etudes Ornithologiques de France Ed. Paris France. 4 compact discs. (can be ordered at SEOF, 55 rue Buffon, 75005 Paris – France)

Odonates and Butterflies

Jacquemin, G. & Boudot, J.-P. 1999. *Les Libellules (Odonates) du Maroc.* Société Française d'Odonatologie Ed., 150 pp. Bois d'Arcy, France. ISBN 2-9507291-3-4 (can be ordered at SFO, 7 rue Lamartine, 78390 Bois d'Arcy – France)

Tannent, J. 1996. *The Butterflies of Morocco, Algeria and Tunisia.* 217 pp.

Mammals

Aulagnier, S. & Thévenot, M. 1986. *Catalogue des Mammifères sauvages du Maroc.* Trav. Inst. Sci., série Zoologie n°41. Rabat.

Bayed, A. & Beaubrun, P. 1987. Les Mammifères marins du Maroc : inventaire préliminaire. *Mammalia* 51: 437-446.

Reptiles

Bons, J. & Geniez, P. 1996. *Amphibiens et Reptiles du Maroc* (with English translation). Asociación Herpetológica Española. Barcelona. 319 pp. Barcelona, Spain. ISBN 84-921999-0-3.

Geniez, P., Mateo, J.-A. & Bons, J. 2000. A checklist of the amphibians and reptiles of Western Sahara (Amphibia, Reptilia). *Herpetozoa,* 13: 149-163.

Schleich, H., Kästle, W. & Kabisch, K. 1996. *Amphibians and Reptiles North Africa: Biology, Systematics, Field Guide.* 630 pp. Koeltz, German

Orchids

Raynaud, C. 1985. *Les Orchidées du Maroc.* Société Française d'Orchidologie Ed. Paris, France. 117 pp. ISBN 2-905734-00-0 (can be ordered at SFO, 84 rue de Grenelle, 75007 Paris, France)

Delforge, P. 2001. *Guide des Orchidées d'Europe, d'Afrique du Nord et du Proche Orient.* Delachaux and Niestlé Eds. Lausanne, Switzerland. 480 pp. ISBN 2-603-01228-2.

Other Flora

Benabid, A. 2000. *Flore et Ecosystèmes du Maroc.* Ibis Press Ed. Paris, France (http://www.ibispress.com/) 360 pp. ISBN 2-910728-13-7.

M'Hirit, O., Benzyane, M., Benchekroun, F., El Yousfi, S.M. & Bendaanoun, M. 1998. *L'Arganier.* Mardaga Ed. Sprimont, Belgium. 150 pp. ISBN 2-87009-684-4.

M'Hirit, O. & Blerot, P. 1999. *Le Grand Livre de la forêt marocaine.* Mardaga Ed. 280 pp. ISBN 2-87009-686-0.

Quézel, P. 2000. *Réflexions sur l'évolution de la flore et de la végétation au Maghreb méditerranéen*. Ibis Press Ed. Paris, France (http://www.ibispress.com/) 118 pp. ISBN 2-910728-15-3.

Travel Guides

A lot of travel guides are available on the market. The following are recommended:

Le Guide du Routard. Maroc. Hachette Ed.
The Rough Guide. Morocco.
Lonely Planet. Morocco

Geography

Coude-Gaussen, G. 1995. *Désertification et aménagement au Maghreb.* L'Harmattan Ed. Paris, France. 314 pp. ISBN 2-73843645-5.

Riser, J. 1996. *Le bassin de Tarfaya (Maroc atlantique, 28°N). Paéloenvironnement, Paléoanthropologie, Préhistoire*. L'Harmattan Ed. Paris, France. ISBN 2-7384-4377-X.

Troin, J.-F. 2002. *Maroc. Régions, pays, territoires*. Tarik Ed., Maisonneuve & Larose.Paris, France. 502 pp. ISBN 2-7068-1630-9.

Mountain – Trekking

Collomb, R.G. 1987. *Atlas Mountain Morocco*. West Col Productions, Goring Reading Berks. 142 pp. ISBN 906227-31-3.

Dresch, J. & Lépiney, J. de 1942 (new edition 1993). *Le Massif du Toubkal*. Belvisi/Edisud Eds. Casablanca, Morocco, and Aix-en-Provence, France. 283 pp. ISBN 9981-819-00-X.

Fougerolles, A. 1991. *Le Haut Atlas. La Montagne des Montagnes*. Glénat Ed. 175 pp. ISBN 2-7234-1272-5.

Knight, R. 2001. *Trekking in the Moroccan Atlas*. Trailblazer publications (www.trailblazer-guides.com). 256 pp. ISBN 1-873756-35-6.

Peyron, M. 1990. Great Atlas traverse. 2 volumes. West Col Productions, Goring Reading Berks. 136 and 134 pp.

Off-road

Gandini, J. 2000. *Pistes du Maroc. Tome I. Haut et Moyen Atlas*. Extrem'Sud Eds. Calvisson, France. 254 pp. ISBN 2-913412-03-3.

Gandini, J. 2001. *Pistes du Maroc. Tome II. Le Sud, du Tafilalet à l'Atlantique.* Extrem'Sud Eds. Calvisson, France. 430 pp. ISBN 2-913412-06-8.

Gandini, J. 2002. *Pistes du Maroc. Tome III. De l'Oued Draa à la Seguiet el Hamra*. Extrem'Sud Eds. Calvisson, France. 236 pp. ISBN 2-913412-07-6.

Archaeology

Collectif 1994. *Milieux, hommes et techniques du Sahara préhistorique.* L'Harmattan Ed. Paris, France. 280 pp.

Rodrigue, A. 1999. *L'art rupestre du Haut Atlas marocain*. L'Harmattan Ed. Paris, France. 420 pp. ISBN 2-7384-8281-3.

'arpets, Rugs and Crafts

Barthélémy, A. 1990. *Tazra. Tapis et Bijoux de Ouarzazate*. Edisud Ed. Aix en Provence, France (http://www.edisud.com). 128 pp. ISBN 2-85744-498-2.

Bouilloc, C., Crouzet, H., Maurières, A. & Vivier, M.F. 2001. *Maroc. Tapis de tribus*. Musée du Tapis et des Arts textiles de Clermond-Ferrand/Edisud. 175 pp. ISBN 2-7449-0155-5.

Maurières, A., Ossart, E. & Lapeyrie, C. 1996. *Au fil du désert. Tentes et tissages des pasteurs nomades de Méditerranée*. Edisud Ed. ISBN 2-85744-806-6.

Ramirez, F. & Rolot, C. 1995. *Tapis et Tissages du Maroc*. ACR Ed. Courbevoie, Paris, France. 192 pp. ISBN 2-86770-085-X.

Samama, Y. 2000. *Le tissage dans le Haut Atlas marocain*. Ibis Press Ed. Paris, France (http://www.ibispress.com/) 96 pp. ISBN 2-910728-11-0.

LOCAL CONTACTS AND SOCIETIES

There are several governmental organisations which deal with nature conservation in Morocco. The 'Ministère de l'Agriculture, du Développement rural et des Eaux et Forêts' (BP 607. Place Abdellah Chefchaouni, Avenue Mohamed V, Quartier Administratif. Rabat, Morocco. http://www.mygale.org/bamouh/madrp.htm) and especially the 'Institut Agronomique et Vétérinaire Hassan II' (BP 620 Madinat Al Irfane. 10101 Rabat, Morocco. http://www.iav.ac.ma) and the wildlife department of 'Administration des Eaux et Forêts' initiate studies on game birds and other hunted or pest species. The 'Centre d'Etude des Migrations d'Oiseaux', a sub-department of the 'Institut Scientifique' (BP 703. Charia Ibn Batouta. 10106 Rabat Agdal, Morocco http://www.israbat.ac.ma) has undertaken studies on the major wetlands and has coordinated mid-winter wildfowl counts since 1989

The GOMAC (Groupe d'Ornithologie du Maroc. c/o Prof. Jacques Franchimont. Quartier Abbass Lemsaadi, rue n°6, n°22. 50000 Meknès VN – Maroc.) is an active NGO bird club; it also co-ordinates the activities of the Moroccan Rare Birds Committee (MRBC). Its journal, *Porphyrio*, publishes annual bird reports and the work results of the MRBC.

To help to improve or update this Guide, please send any comment sightings or reports from birdwatching trips to Morocco to:

Patrick Bergier
4 Avenue Folco de Baroncelli
13210 – Saint Rémy de Provence
France

website: http://www.ifrance.com/go-south
e-mail: pbergier@yahoo.fr